70 & 70 Gravitation Fest

Journal of Physics: Conference Series Volume 831

Cartagena, Columbia
28 – 30 September 2016

ISBN: 978-1-5108-3987-8
ISSN: 1742-6588

Printed from e-media with permission by:

Curran Associates, Inc.
57 Morehouse Lane
Red Hook, NY 12571

Some format issues inherent in the e-media version may also appear in this print version.

Printed by Curran Associates, Inc. (2017)

For permission requests, please contact the Institute of Physics
at the address below.

Institute of Physics
Dirac House, Temple Back
Bristol BS1 6BE UK

Phone: 44 1 17 929 7481
Fax: 44 1 17 920 0979

techtracking@iop.org

Additional copies of this publication are available from:

Curran Associates, Inc.
57 Morehouse Lane
Red Hook, NY 12571 USA
Phone: 845-758-0400
Fax: 845-758-2633
Email: curran@proceedings.com
Web: www.proceedings.com

Table of contents

Volume 831

70 & 70 Gravitation Fest
28–30 September 2016, Cartagena, Colombia

Accepted papers received: 23 March 2017
Published online: 7 April 2017

Preface

Papers

IOP Conf. Series: Journal of Physics: Conf. Series **831** (2017) 011001 doi:10.1088/1742-6596/831/1/011001

A preface to the 70&70 Gravity Fest

A. C. Gutiérrez-Piñeres
Facultad de Ciencias Básicas, Universidad Tecnológica de Bolívar, Cartagena, Colombia
E-mail: acgutierrez@unitecnologica.edu.co

E. A. Montoya
Escuela de Física, Universidad Industrial de Santander, Bucaramanga, Colombia
E-mail: emontoya@uis.edu.co

L.A. Núñez
Escuela de Física, Universidad Industrial de Santander, Bucaramanga-Colombia and
Departamento de Física, Universidad de Los Andes, Mérida Venezuela.
E-mail: lnunez@uis.edu.co

There are transcendental landmarks in life to account and balance achievements and failures. With no doubt the age of 70 brings with it the possibility to trace back a significant part of our existences and can considered a natural stop. The year 2016 is, for many Latin American physicists, an important date because Rodolfo Gambini Italiano from the Universidad de la Repblica, Uruguay and Luis (Gaucho) Herrera Cometta from Universidad Central de Venezuela reach this age enjoying the passion for building generations of physicis in Latin America and leaving a profound and important imprint in Theoretical Physics along several countries of our continent. This is a tribute to these two masters, a great opportunity to share their visions and particular stiles with new generations of physics students from Latin America.

These two masters of the Theoretical Physics have life-lines that crosses at different instants of their lives. Rodolfo (born May 11, 1946), got degree in physics from the University of the Republic in 1972–In fact, he was the first physicist graduated in Uruguay–. Then, he moved to Paris and obtained his doctorate in Theoretical Physics under the supervision of Achilles Papapetrou. This was the first encounter with El Gaucho. Luis (born December 20, 1946) after graduated from the People's Friendship University, Moscow went to Paris to study also under the guidance of Papapetrou. Luis returned to Caracas in 1972, taking a position at the Universidad Central de Venezuela and joint Professor Carlos Aragone and Gustavo González Martín, starting a seminal group the Caracas Relativity Seminar. Prof. Aragone, former supervisor of Rodolfo, had just arrived to Caracas after militar *coup d'état* in Uruguay and got a position at Universidad Simón Bolívar, USB. In 1975 Rodolfo also started to work at USB, invited by Prof Aragone and this is the second cross road between him and El Gaucho. In 1978 Rodolfo and Antoni Trias began a great friendship and a fruitful collaboration which leads to the invention of the loop representation for Yang-Mills theories in 1986.

Luis Gaucho Herrera is a versatile theoretical physicist with major contribution on the effect of anisotropy on the evolution of selfgravitating compact objects. He also pioneer in the heritage of symmetries within General Relativity and in the application of Extended Thermodynamics for Astrophysical scenarios, where he has very deep and important contributions isolating dissipation effects on relativistic systems. Recently, he has proposed alternative approaches to detect gravitational radiation using gyroscopes and about the relevance of super energy and super Poynting in General Relativity.

Rodolfo Gambini is a profound physicist with contributions ranging from philosophy of science and foundations of quantum mechanics to lattice gauge theories to quantum gravity. He has also an active participation building public scientific policies in Uruguay and in recent years he has studied issues in the foundations of quantum mechanics, having developed the Montevideo Interpretation of Quantum Mechanics. He also found the exact solution of the quantum Einstein equations in loop quantum gravity for vacuum spherically symmetric space-times, which resolves the singularity inside black holes.

Thanks to Dean of Facultad Ciencias Básicas Dr. Jorge Luis Muñiz and the hospitality of Ingrid Tamayo and Rosana Ávila members of the Educación Permanente staff of Universidad Tecnológica de Bolívar (Cartagena-Colombia) we met from 28th to 30th of September 2016.

Among the researchers that have the opportunity to share visions with Rodolfo and Luis about Classical and Quantum Gravity we can mention: Iván Agullo (Louisiana State University, USA.), Miguel Alcubierre (Universidad Nacional Autónoma de México, México), Roberto Aquilano (Universidad de Rosario Argentina), Jaume Carot (Universitat de les Illes Balears, Spain), Alicia Di Prisco (Universidad Central de Venezuela, Venezuela), Guillermo González (Universidad Industrial de Santander, Bucaramanga, Colomba) César López Monsalvo (Universidad Autónoma Metropolitana, México), Justo Ospino (Universidad de Salamanca, Spain), Jorge Pullin (Louisiana State University, USA.), Hernando Quevedo (Universidad Nacional Autónoma de México, México), Alvaro Restuccia (Universidad de Antofagasta, Chile), Yeinzon Rodríguez (Universidad Industrial de Santander, y Universidad Antonio Nariño, Colombia), Olivier Sarbach (Universidad Michoacana de San Nicolás de Hidalgo, México) and Roberto Sussman (Universidad Nacional Autónoma de México, México).

1. Acknowledgment

We gratefully thanks for financial support to Universidad Tecnológica de Bolívar, Vicerrectoría de Investigación from Universidad Industrial de Santander and Patrimonio Autónomo fondo nacional de financiamiento para la ciencia, la tecnología y la innovación, Francisco José de Caldas, which make possible this meeting.

IOP Conf. Series: Journal of Physics: Conf. Series **831** (2017) 011001 doi:10.1088/1742-6596/831/1/011001

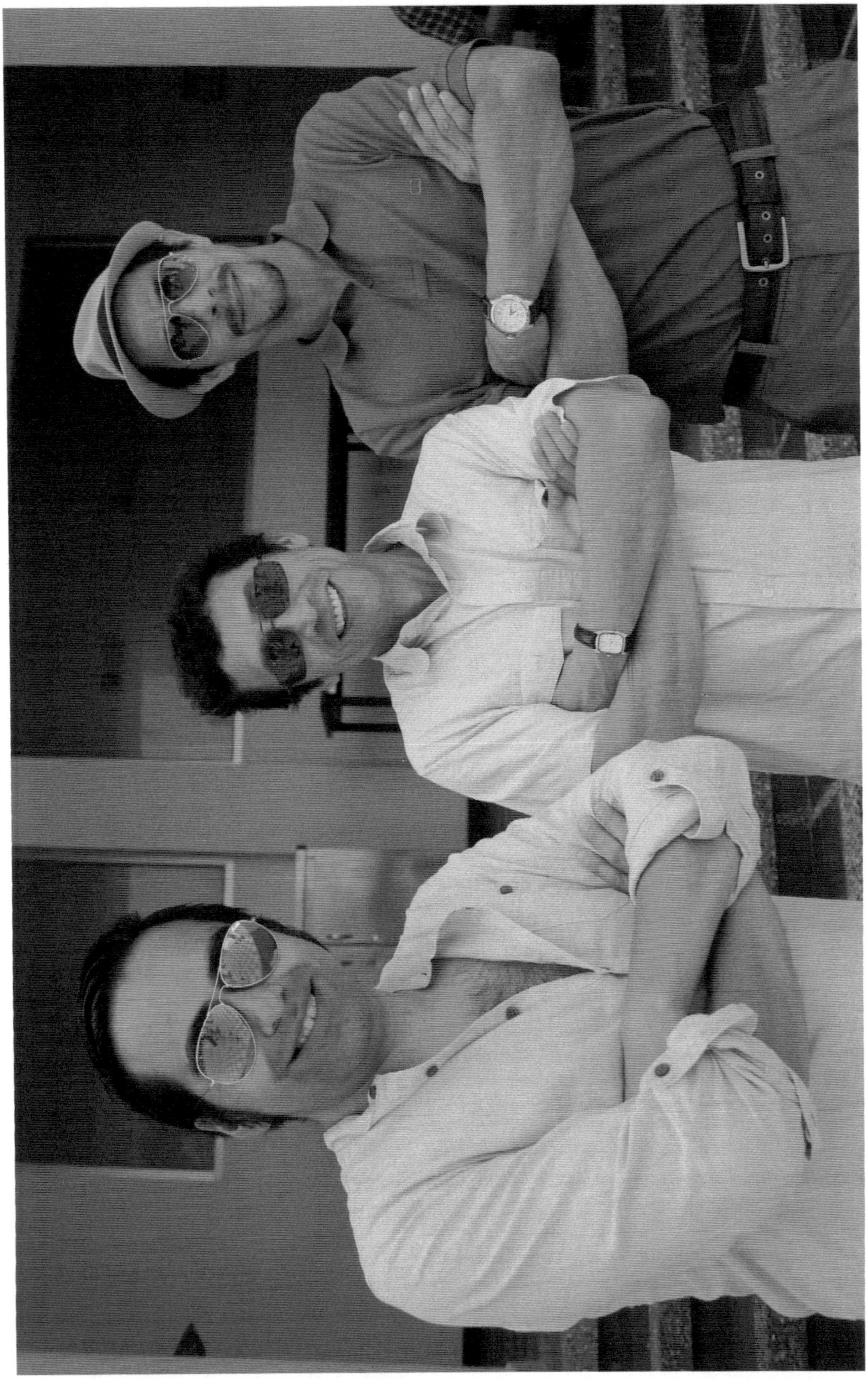

IOP Conf. Series: Journal of Physics: Conf. Series **831** (2017) 011002 doi:10.1088/1742-6596/831/1/011002

Peer review statement

All papers published in this volume of *Journal of Physics: Conference Series* have been peer reviewed through processes administered by the proceedings Editors. Reviews were conducted by expert referees to the professional and scientific standards expected of a proceedings journal published by IOP Publishing.

IOP Conf. Series: Journal of Physics: Conf. Series **831** (2017) 012001 doi:10.1088/1742-6596/831/1/012001

The Theory of Gravitation: A Tale of Many Questions and Few Answers

Herrera L
Instituto Universitario de Física Fundamental y Matemáticas, Universidad de Salamanca, Salamanca 37007, Spain

E-mail: lherrera@usal.es

Abstract. We discuss on different issues pertaining the theory of gravity, which pose some unresolved fundamental questions. First we tackle the problem of observers in general realtivity, with particular emphasis in tilted observers. We explain why these observers may detect dissipative processes in systems which appear isentropic to comoving observers. Next we analyze the strange relationship between vorticity and radiation, and underline the potential observational consequences of such a link. Finally we summarize all the results that have been obtained so far on the physical properties of the sources of gravitational radiation. We conclude with a list of open questions which we believe deserve further attention.

1. Introduction

"There is something fascinating about science. One gets such wholesale returns of conjecture out of such a trifling investment of fact – Mark Twain

In this manuscript I would like to elaborate on some issues that I raised recently, at the occasion of the "70&70 Gravitation Fest", which took place in Cartagena, Colombia on September 28-30, 2016, to celebrate the 70th birthday of Rodolfo Gambini and myself.

The text is heavily based on the conference I delivered at that occasion, but contains some new aspects of the analyzed problems, which were absent in my conference.

I shall focus on three issues, recurrently appearing in the theory of gravity (in any theory of gravity, not only general relativity), which seem to me particularly appealing and endowed with potential for further development. These are:

- What is the role played by the observers in general relativity? What can we learn from the study of tilted spacetimes? How observer dependent is the concept of irreversibility?
- What is the link between vorticity and gravitational radiation? Is it a cause–effect relationship? Or are they concomitant? What are the possible observational consequences of such a link?
- How to relate the gravitational radiation with the physical properties of its source? What constraints should be impossed on the source to supress the emission of gravitational radiation? What is the spacetime outside a source of gravitational radiation? How does the emission of gravitational radiation start (as seen from the source)? How does this process cease? Are there wave tails?

2. Tilted and comoving observers, and the definition of irreversibility (entropy)

"Irreversibility is a consequence of the explicit introduction of ignorance into the fundamental laws." M. Born

In general relativity there exist an ambiguity in the description of the source, related to the arbitrariness in the choice of the four-velocity in terms of which the energy momentum tensor is split. Thus, for example, one possible interpretation of a given spacetime may correspond to a congruence of observers comoving with the fluid, whereas the other corresponds to the observer congruence which has been (Lorentz) boosted, with respect to the former. In such a case, both, the general properties of the source and the kinematic properties of the congruence would be different. Many examples of this kind have been analyzed in the past (see for example [1]–[15] and references therein)

For example, in the case of the zero curvature FRW model, we have a perfect fluid solution for observers at rest with respect to the timelike congruence defined by the eigenvectors of the Ricci tensor, whereas for observers moving relative to the previously mentioned congruence of observers, it can also be interpreted as the exact solution for a viscous dissipative fluid [3]. It is worth noticing that the relative ("tilting") velocity between the two congruences may be related to a physical phenomenon such as the observed motion of our galaxy relative to the microwave background radiation [7].

Thus, zero curvature FRW models as described by "tilted" observers will detect a dissipative fluid and energy–density inhomogeneity, as well as different values for the expansion scalar and the shear tensor, among other differences, with respect to the "standard" (comoving) observers (see [3] for a comprehensive discussion on this example).

Now, the question arises: is the heat flux vector observed by tilted observers, associated to irreversible processes or not? The rationale behind such a question resides in the fact that in the past some authors have argued that dissipative fluids (understood as fluids whose energy–momentum tensors present a non–vanishing heat flux contribution), are not necessarily incompatible with reversible processes (e.g see [16]–[18]).

More specifically, in the context of the standard Eckart theory [19], a necessary condition for the compatibility of an imperfect fluid with vanishing entropy production (in the absence of bulk viscosity) is the existence of a conformal Killing vector field CKV) χ^α such that $\chi^\alpha = \frac{V^\alpha}{T}$ where V^α is the four–velocity of the fluid and T denotes the temperature. In the context of causal dissipative theories, e.g. [20]–[25], the existence of such CKV is also necessary for an imperfect fluid to be compatible with vanishing entropy production (see [8]).

However, such a claim should not worry us, for two reasons. On the one hand, a carefull analysis of the problem readily shows, that the compatibility of reversible processes with the existence of dissipative fluxes becomes trivial if a constitutive transport equation is adopted. Indeed, in this latter case such compatibility forces the heat flux vector to vanish as well. In other words, even if a non–vanishing heat flow vector is assumed to exist, the imposition of the CKV and the vanishing entropy production condition, cancel the heat flux, once a transport equation is assumed (see [26] for a detailed discussion on this point). In other words, in the presence of a CKV of the kind mentioned before, the assumption of a transport equation whether in the context of the Eckart–Landau theory, or a causal theory, implies that a vanishing entropy production leads to a vanishing heat flux vector. Therefore, under the conditions above, the system is not only reversible but also non dissipative.

On the other hand, neither LTB [27, 28, 29] nor the Szekeres spacetimes [30, 31] admit a CKV, accordingly we may safely conclude that the heat flux vector appearing in these cases (at least), is associated to truly (entropy producing) dissipative processes [8, 9].

Thus, an intriguing question arises, namely: how is it possible that tilted observers may detect irreversible processes, whereas comoving observers describe an isentropic situation ?

IOP Conf. Series: Journal of Physics: Conf. Series **831** (2017) 012001 doi:10.1088/1742-6596/831/1/012001

As we shall see below, the answer to the above question is closely related to definition of entropy, which is highly observer dependent, as illustrated, for example, by the Gibbs paradox.

Indeed, entropy is a measure of how much is not known (uncertainty). The fact that physical objects do not have an intrinsic uncertainty (entropy) has been illustrated in great detail in [32, 33].

The "antropomorphic" nature of the concept of entropy makes itself evident in the Gibbs paradox. In its simplest form, the paradox appears from the consideration of a box divided by a wall in two identical parts, each of which is filled with an ideal gas (at the same pressure and temperature). Then if the partition wall is removed, the gases of both parts of the box will mix.

If the gases from both sides are distinguishable, the entropy of the system will rise, while there is no increase in entropy if they are identical. This leads to the striking conclusion that irreversibility (and thereby entropy), depends on the ability of the observer to distinguish, or not, the gases from both sides of the box. In other words, irreversibility would depend on our knowledge of physics [34], confirming thereby our previous statement that physical objects are deprived of intrinsic entropy. It can only be defined **after** the number of states that are accesible by the system, is established.

Now, if a given physical system is studied by a congruence of comoving observers, then the three–velocity of any given fluid element is automatically assumed to vanish, whereas for the tilted observers this variable represents an additional degree of freedom. Therefore, the number of possible states in the latter case is much larger than in the former one.

In orther to obtain the tilted congruence (from the comoving one), we have to submit locally Minkowskian comoving observers to a Lorentz boost. Such an operation, of course, is performed locally, and no global transformation exists linking both congruences. The point is that, passing from the tilted congruence to the comoving one, we usually overlook the fact that both congruences of observers store different amounts of information. Here resides the clue to resolve the quandary mentioned above, about the presence or not of dissipative processes, depending on the congruence of observers that carry out the analysis of the system.

Thus, since for the comoving observers the system is dissipationless, it is clear that the increasing of entropy, when passing to the tilted congruence, should imply the presence of dissipative (entropy producing) fluxes, in the latter.

It is instructive to take a look on this issue from a different perspective. Thus, let us consider the transition from the tilted congruence to the comoving one. According to the Landauer principle, [35] (also referred to as the Brillouin principle [36]–[40]), the erasure of one bit of information stored in a system requires the dissipation into the environment of a minimal amount of energy, whose lower bound is given by

$$\triangle E = kT \ln 2, \tag{1}$$

where k is the Boltzmann constant and T denotes the temperature of the environment.

In the theory of information, erasure, is just a reset operation restoring the system to a specific state, and is achieved by means of an external agent. In other words, one can decrease the entropy of the system by doing work on it, but then one has to increase the entropy of another system (or the environment). So to speak, Landauer principle is an expression of the fact that logical irreversibility necessarily implies thermodynamical irreversibility.

Thus, transforming to the comoving congruence, we reset the value of the three–velocity (of any fluid element) to zero, which implies that the information has been erased, and a decrease of entropy occurs, but we have not any external agent (since we are considering self–gravitating systems), and therefore such a decrease of entropy is accounted by the dissipative flux observed in the tilted congruence (remember that in the comoving congruence the system is dissipationless).

We may summarize the main issues addressed in this section, in the following points:

- Uncertainty (entropy) is highly dependent on the observer.

- Comoving and tilted observers, store different amounts of information.
- According to the Landauer principle, erasure of information is always accompanied by dissipation (you have to pay to forgetting).
- The quandary mentioned above, is resolved at the light of the three previous comments.

3. Radiation and vorticity

"How often have I said to you that when you have eliminated the impossible, whatever remains, however improbable, must be the truth?"- Sherlock Holmes in: The sign of four.

The first theoretical evidence relating gravitational radiation to vorticity, appeared when it was established that in an expansion of inverse powers of r (where r denotes the null Bondi coordinate [41]), the coefficient of the vorticity at order $1/r$ will vanish if and only if there are no news (no gravitational radiation) [42], implying that a frame dragging effect is associated with gravitational radiation. This result was further confirmed in [43]–[46].

The intriguing fact is that these two (very different) phenomena are related. More specifically, we wonder if there is any physical reason behind the link between vorticity and radiation.

The hint to solve this quandary comes from an idea put forward by Bonnor in order to explain the appearance of vorticity in the spacetime generated by a charged magnetic dipole [47].

For such a system, as was observed by Bonnor, there exists a non–vanishing component of the Poynting vector, describing a flow of electromagnetic energy round in circles [48]. He then suggested that such a circular flow of energy affects inertial frames by producing vorticity of congruences of particles, relative to the compass of inertia. This conjecture has been shown to be valid for a general axially symmetric stationary electrovacuum metric [49]. It is worth mentioning that this "circular" flow of electromagnetic energy is endowed with a solid physical meaning, since it is absolutely necessary in order to preserve the conservation of angular momentum, in the well known "paradox" of the rotating disk with charges and a solenoid [48].

Then it came to our minds [45], the idea that a similar mechanism might be at the origin of vorticity in the gravitational case, i.e. that a circular flow of gravitational energy would produce vorticity. However the nonexistence of a local and invariant definition of gravitational energy, rose at that time the question about what expression for the "gravitational" Poynting vector should be used. In [46] we tried with the super–Poynting vector based on the Bel–Robinson tensor [50]–[52].

Doing so, we succeded to establish the link between gravitational radiation and vorticity, invoking a mechanism similar to that proposed by Bonnor for the charged magnetic dipole. It was shown that the resulting vorticity is always associated to a circular flow of superenergy on the plane orthogonal to the vorticity vector. It was later shown that the vorticity appearing in stationary vacuum spacetimes also depends on the existence of a flow of superenergy on the plane orthogonal to the vorticity vector [53]. Furthermore in [54] it was shown that not only gravitational but also electromagnetic radiation produces vorticity. In this latter case we were able to isolate contributions from, both, the electromagnetic Poynting vector as well as from the super–Poynting vector.

It is worth noticing that for unbounded configurations (e.g. cylindrically symmetric sytems), radiation does not produce vorticity if there are no circular flows of superenergy on any plane of the 3-space (e.g. Einstein–Rosen spacetime), which reinforces further the role of the circular flow of super–energy as the link between radiation and vorticity.

Besides the evident theoretical interest of the issue discused in this section, we would like to emphasize the very important potential observational consequences of the association of vorticity with radiation. Indeed, the direct experimental evidence of the existence of the Lense–Thirring effect [55]–[57] brings out the high degree of development achieved in the technology required to measure rotations. In the same direction point recent proposals to detect frame dragging by

IOP Conf. Series: Journal of Physics: Conf. Series **831** (2017) 012001 doi:10.1088/1742-6596/831/1/012001

means of ring lasers [58]–[62]. Also it is worth mentioning the possible use of atom interferometers [63]– [65], atom lasers [66], anomalous spin–precession experiments [67] and matter wave Sagnac interferometers [68], to measure vorticity.

Le us close this section with a remark about an important feature.

At order $(1/r^2)$ there are contributions to the vorticity with a time dependent term not involving news (i.e. not associated with gravitational radiation). This last term represents the class of non-radiative motions discussed by Bondi [41] and may be thought to correspond to the tail of the wave, appearing after the radiation process has ended [69]. Thus, the obtained expression allows for "measuring" (in a gedanken experiment, at least) the wave-tail field. This in turn implies that observing the gyroscope, for a period of time from an initial static situation until after the vanishing of the news, should allow for an unambiguous identification of a gravitational radiation process.

However, as it has been recently shown (see the subsection 4.6 below) the transition from a state of radiation (gravitational) to an equilibrium state, is not forbidden, after a small time interval of the order of magnitude of the hydrostatic time, the relaxation time, and the thermal adjustment time. This result is in contradiction with previous results [41], [69]–[76], that suggest that such a transition is forbidden, due to the appearance of the wave tails mentioned above.

Therefore the detection (or not) of the wave tail, would have very important theoretical consequences.

4. Gravitational radiation and its source

"I have no data yet. It is a capital mistake to theorize before one has data. Insensibly one begins to twist facts to suit theories, instead of theories to suit facts".–Sherlock Holmes in: A scandal in Bohemia.

In the sixties, there was a blossom of very powerful methods to study gravitational radiation, beyond the well known linear approximation, [41, 77, 78, 79]. All of them focused on the behaviour of the field very far from the source, whose specific properties did not enter into the discussion, to avoid the appearance of caustics and similar pathologies.

Besides the many fundamental results obtained from these methods, their main merit consists in including non–linear effects, which are known to play a very important role in general relativity. The absence of a detailed discussion on the behaviour of the sources of gravitational radiation, was the triggering motivation to undertake the task to develop a general formalism that could provide a description of the effects of gravitational radiation on the physical properties of the source, and viceversa.

In other words we searched to establish the relationship between gravitational radiation and source properties.

The fact that we want to describe gravitational radiation, force us to depart from the spherical symmetry. On the other hand, since we are willing to provide an analytical description of the problem, avoiding as much as possible to resort to numerical methods, we need to impose additional restrictions. Thus, we shall rule out cylindrical symmetry on physical grounds, but we shall assume axial and reflection symmetry, which as shown in [80] is the highest degree of symmetry of the Bondi metric [41], which does not prevent the emission of gravitational radiation.

The general formalism mentioned before was presented in [81]. It was obtained by using the 1+3 formalism [82, 83, 84], in a given coordinate system, and resorting to a set of scalar functions known as Structure Scalars [85], which have been shown to be very useful in the description of self–gravitating systems [86]–[95].

A detailed description of this general formalism may be found in [81], here we shall focus on the main results obtained so far on this issue, and the open questions which we believe deserve further attention.

4.1. The shear–free case

We started by analyzying the shear free case [96]. Under such a condition we have found that for a general dissipative and anisotropic fluid, vanishing vorticity, is a necessary and sufficient condition for the magnetic part of the Weyl tensor to vanish, thereby providing a generalization of the same result for perfect fluids obtained in [97, 98, 99]. This result, in turn, implies that vorticity should necessarily appear if the system radiates gravitationally. We stress that this result is not restricted to the axially (and reflection) symmetric case. This further reinforces the well established link between radiation and vorticity discussed in the previous section.

If besides the shear–free condition, we assume further that the fluid is geodesic, then the vorticity vanishes (and thereoff the magnetic part of the Weyl tensor) and no gravitational radiation is produced. In this case our study generalizes the models studied by Coley and McMannus [100, 101]. A similar result is obtained for the cylindrically symmetric case [102], suggesting a link between the shear of the source and the generation of gravitational radiation.

Finally, in the geodesic case, we also observe that, in the non-dissipative case, the models do not need to be FRW (as already stressed in [100]), however the system tends to the FRW spacetime (if Θ is positive). In presence of dissipative fluxes, such a tendency does not appear, which illustrates further how relevant, the dissipative processes may be.

4.2. The perfect, geodesic fluid

Next we have considered the the restricted case, when the fluid is perfect and geodesic, without assuming *ab initio* the shear–free condition [103]. As the result of such study we have found that: All possible models compatible with our metric, and the perfect fluid plus the geodesic condition, are FRW, and accordingly non–radiating (gravitationally). It seems that, both, the geodesic and the non–dissipative conditions, are quite restrictive, when looking for a source of gravitational waves.

It is worth noticing that, not only in the case of dust, but also in the absence of dissipation in a perfect fluid, the system is not expected to radiate (gravitationally) due to the reversibility of the equation of state. Indeed, radiation is an irreversible process, a fact that emerges at once if absorption is taken into account and/or Sommerfeld type conditions, which eliminate inward traveling waves, are imposed. Therefore, the irreversibility of the process of emission of gravitational waves, must be reflected in the equation of state through an entropy increasing (dissipative) factor.

We should recall that geodesic fluids not belonging to the class considered here (Szekeres) have also been shown not to produce gravitational radiation. This strengthens further the case of the non–radiative character of pure dust distributions.

4.3. The dissipative, geodesic fluid

Since dissipation should be present in any scenario where gravitational radition is produced, we decided to study the simplest possible fluid distribution which we might expect to be compatible with a gravitational radiation, i.e. a dissipative dust under the geodesic condition [104].

Two possible subcases were considered separately, namely: the fluid distribution is assumed, from the beginning, to be vorticity–free, or not.

In the former case, it was shown that the vanishing of the vorticity implies the vanishing of the heat flux vector, and therefore, the resulting spacetime is FRW.

In the latter case, it is shown that the enforcement of the regularity conditions at the center, implies the vanishing of the dissipative flux, leading also to a FRW spacetime.

Thus all possible models, sourced by a dissipative geodesic dust fluid, belonging to the family of the line element considered here, do not radiate gravitational waves during their evolution, unless regularity conditions at the center of the distribution are relaxed.

In other words physically acceptable models require the inclusion of, both, dissipative and anisotropic stresses terms, i.e. the geodesic condition must be abandoned. In this case, purely analytical methods are unlikely to be sufficient to arrive at a full description of the source, and one has to resort to numerical methods.

4.4. The space–time outside the source of gravitational radiation
Based on the fact that the process of gravitational radiation is an irreversible one, and therefore must entail dissipative processes within the source, we should conclude that there should be an incoherent radiation (null fluid) at the outside of the source, produced by those dissipative processes. Keeping this fact in mind, we should remark that the Bondi–Sachs metric [41], [77], should be regarded as an approximation to the space–time outside the source, when the null fluid produced by the dissipative processes is neglected.

Starting with the description of this null fluid, we apply the formalism developped in [81], to study some of the properties of such a null fluid [105].

As the main result of our study we found that the absence of vorticity implies that the exterior spacetime is either static or spherically symmetric (Vaidya). Reinforcing thereby the fundamental role of vorticity in any process involving production of gravitational radiation, already stressed.

The spherically symmetric case (Vaidya) was, asymptotically, recovered within the context of the 1+3 formalism [105].

4.5. Leaving the equilibrium
Next, as an application of our general method [81], we analyzed the situation, just after its departure from hydrostatic and thermal equilibrium, at the smallest time scale at which the first signs of dynamic evolution appear [106]. Such a time scale is smaller than the thermal relaxation time, the thermal adjustment time and the hydrostatic time. Specifically we were able to answer to the following questions: 1) what are the first signs of non–equilibrium? 2) which physical variables do exhibit such signs? 3) what does control the onset of the dynamic regime, from an equilibrium initial configuration?

It was obtained that the onset of non–equilibrium will critically depend on a single function directly related to the time derivative of the vorticity. Among all fluid variables (at the time scale under consideration), only the tetrad component of the anisotropic tensor, in the subspace orthogonal to the four–velocity and the Killing vector of axial symmetry, shows signs of dynamic evolution. Also, the first step towards a dissipative regime begins with a non–vanishing time derivative of the heat flux component along the meridional direction. The magnetic part of the Weyl tensor vanishes (not so its time derivative), indicating that the emission of gravitational radiation will occur at later times. Finally, the decreasing of the effective inertial mass density, associated to thermal effects, was clearly illustrated [106].

4.6. Reaching the equilibrium
Next, we described the transition of a gravitationally radiating, axially and reflection symmetric dissipative fluid, to a non–radiating state [107]. What we wanted to elucidate was if, very shortly after the end of the radiating regime, at a time scale of the order of the thermal relaxation time, the thermal adjustment time and the hydrostatic time (whichever is larger), the system reaches the equilibrium state. We recall that in all the studies carried out in the past, on gravitational radiation outside the source, such a transition to a static case, is forbidden. However, neither of these studies include the physical properties of the source, giving rise the possibility of a quite different result.

Using the general method presented in [81], we were able to prove that the system does in fact reach the equilibrium state. This result is at variance with previous results mentioned above, implying that such a transition is forbidden. The reason for such a discrepancy resides in the fact that some elementary, but essential, physical properties of the source, have been overlooked in these latter studies. Our result strengths further the relevance of the physical properties of the source, in any discussion about the physical properties of the field. Also, it emphasizes the need to resort to global solutions, whenever important aspects about the behaviour of the gravitational field are discussed. In other words, the coupling between the source and the external field may introduce important constraints on the physical behaviour of the system, implying that details of the source fluid cannot be left out, because they may be relevant to distant GW scattering.

4.7. The quasi–static regime

As it is well known, in the study of self–gravitating fluids we may consider three different possible regimes of evolution, namely: the static, the quasi–static and the dynamic.

In the static case, the spacetime admits a timelike, hypersurface orthogonal, Killing vector. Thus, a coordinate system can always be choosen, such that all metric and physical variables are independent on the time like coordinate. The static case, for axially and reflection symmetric spacetimes, was studied in [108].

Next, we have the full dynamic case where the system is considered to be out of equilibrium (thermal and dynamic), the general formalism to analyze this situation, for axially and reflection symmetric spacetimes was developed in [81].

In between the two regimes described above, we have the quasi–static evolution.

In this regime the system is assumed to evolve, although sufficiently slow, so that it can be considered to be in equilibrium at each moment. This means that the system changes slowly, on a time scale that is very long compared to the typical time in which the fluid reacts to a slight perturbation of hydrostatic equilibrium. This typical time scale is called hydrostatic time scale (sometimes this time scale is also referred to as dynamical time scale).

Thus, in this regime the system is always very close to hydrostatic equilibrium and its evolution may be described as a sequence of equilibrium configurations.

In other words, we may assume safely the quasi–static approximation (QSA) whenever all the relevant characteristic times of the system under consideration, are much larger than the hydrostatic time.

This assumption is very sensible because the hydrostatic time scale is very small for many phases of the life of the star. It is of the order of 27 minutes for the Sun, 4.5 seconds for a white dwarf and 10^{-4} seconds for a neutron star of one solar mass and 10 Km radius. It is a fact that any of the stellar configurations mentioned above, generally (but not always), changes on a time scale that is very long compared to their respective hydrostatic time scales.

Motivated by the comments above, we applied the framework developped in [81], to carry out a study of axially and reflection symmetric fluids in the quasi–static regime [109].

For doing that we needed to introduce different invariantly defined "velocities", in terms of which the QSA is expressed.

It was obtained that the shear and the vorticity of the fluid, as well as the dissipative fluxes, may affect the (slow) evolution of the configuration, as to produce "splittings" within the fluid distribution.

Also it was shown that in the QSA, the contributions of the gravitational radiation to the components of the super–Poynting vector do not necessarily vanish. However, it appears that if at any given time, the magnetic part of the Weyl tensor vanishes, then it vanishes at any other time afterwards. Thus it is not reasonable to expect gravitational radiation from a physically meaningful system, radiating for a finite period of time (in a given time interval) in the QSA.

5. OPEN ISSUES

As the reader should easily understand, the three issues considered here, still present a great deal of unanswered questions. Below we display a partial list of problems which we believe deserve some attention:

- How could one describe the "cracking" (splitting) of the configurations, in the context of this formalism ? How the appearance of such crackings would affect the emission of gravitational radiation?
- We do not have an exact solution (written down in closed analytical form) describing gravitational radiation in vacuum, from bounded sources. Accordingly, any specific modeling of a source, and its matching to an exterior, should be done numerically. How such a matching could be done?
- It should be useful to introduce the concept of the mass function, similar to the one existing in the spherically symmetric case (an extension of the Bondi mass for the interior of the source). This could be relevant, in particular, in the matching of the source to a specific exterior.
- How does the system analyzed in [81] look for tilted observers? Is it possible that tilted observers detect gravitational radiation from systems that are non–radiating (gravitationally), as seen by comoving observers?
- What could we learn by imposing different kind of symmetries (more general that isometries) on the axially symmetric dissipative fluids studied in [81].
- What is the threshold of sensitivity in the measure of vorticity, reached by the present technology? What are the expected values for the potential future developments of different experiments aiming the detection of rotations? Are these values within the range expected for realistic sources of gravitational radiation?

Acknowledgments

It is a real pleasure to thanks the organizying committee of "70&70 Gravitation Fest": Antonio C. Gutiérrez-Piñeres (Universidad Tecnológica de Bolívar); Edison Montoya (Universidad Industrial de Santander); Jorge Muñiz (Universidad Tecnológica de Bolívar) and Luis A. Núñez (Universidad Industrial de Santander), for their generosity and the great effort deployed to make the event possible.

References

[1] King A R and Ellis G F R 1973 *Commun. Math. Phys.* **31**, 209
[2] Tupper B O J 1981 *J. Math. Phys.* **22**, 2666
[3] Coley A A and Tupper B O J 1983 *Astrophys. J* **271**, 1
[4] Coley A A and Tupper B O J 1983 *Gen. Rel. Grav.* **15**, 977
[5] Tupper B O J 1983 *Gen. Rel. Grav.* **15**, 849
[6] Coley A A and Tupper B O J 1984 *Phy. Lett. A* **100**, 495
[7] Coley A A 1987 *Astrophys. J.* **318**, 487
[8] Herrera L, Di Prisco A and Ibáñez J 2011 *Phys. Rev. D* **84**, 064036
[9] Herrera L, Di Prisco A, Ibáñez J and Carot J 2012 *Phys. Rev. D* **86**, 044003
[10] Tsagas C G and Kadiltzoglou M I 2013 *Phys. Rev. D* **88**, 083501
[11] Kumar R and S. Srivastava 2013 *Astrophys. Space Sci.* **346**, 567
[12] Sharif M and Tahir H 2014 *Astrophys. Space Sci.* **351**, 619
[13] Fernández J and Pascual–Sánchez J 2014 *Procc. Math. Stat.* **60**, 361
[14] Sharif M and Zaeem Ul Haq Bhatti M 2015 *Int. J. Mod. Phys. D* **24**, 1550014
[15] Yousaf Z, Bamba K and Zaeem Ul Haq Bhatti M 2017 *Phys. Rev. D* **95**, 024024
[16] Stephani H 1982 *Introduction to General Relativity* (Cambridge: Cambridge University Press)
[17] Bedran M L and Calvao M O 1993 *Class. Quantum Grav.* **10**, 767
[18] Triginer J and Pavón D 1995 *Class. Quantum. Grav.* **12**, 199

[19] Eckart C 1940 *Phys. Rev.* **58**, 919
[20] Müller I 1967 *Z. Physik* **198**, 329
[21] Israel W 1976 *Ann. Phys.* (NY) **100**, 310
[22] Israel W and Stewart J 1976 *Phys. Lett. A* **58** 213
[23] Israel W and Stewart J 1979 *Ann. Phys.* (NY) **118**, 341
[24] Jou D, Casas-Vázquez J and Lebon G 1988 *Rep. Prog. Phys.* **51**, 1105
[25] Jou D, Casas-Vázquez J and Lebon G 1993 *Extended irreversible Thermodynamics* (Berlin: Springer)
[26] Herrera L, Di Prisco A and Ibáñez J , 2012 *Phys. Lett. A* **376**, 899
[27] Lemaître G 1933 *Ann. Soc. Sci. Bruxelles* **A 53**, 51 and re-printed Lemaître G 1997 *Gen. Rel. Grav.* **29**, 641
[28] Tolman R C 1934 *Proc. Natl. Acad Sci* **20**, 169 and re-printed Tolman R C 1997 *Gen. Rel. Grav.* **29**, 935
[29] Bondi H 1947 *Mon. Not. R. Astron. Soc.* **107**, 410 and re-printed Bondi H 1999 *Gen. Rel. Grav.* **31**, 1783
[30] Szekeres P 1975 *Phys. Rev. D* **12**, 2941
[31] Szekeres P 1975 *Commun. Math. Phys.* **41**, 55
[32] Adami C 2016 *Phil. Trans. Roy. Soc. A* **374**, 20150230
[33] Jaynes E T 1965 *Amer. J. Phys.* **33**, 391
[34] Bais F A and Farmer J D 2007 *Preprint physics.class-ph 0708.2837v2.*
[35] Landauer R 1961 *IBM Res. Develop.* **5**, 183
[36] Brillouin L 1953 *J. Appl. Phys.* **24**, 1152
[37] Brillouin L 1964 *Scientific Uncertainty and Information* (New York: Academic)
[38] Brillouin L 1962 *Science and Information Theory* (New York: Academic)
[39] Kish L B and Granqvist C G 2011 *Preprint physics.gen-ph 1110.0197v7.*
[40] Kish L B 2004 *Proc. IEE* **151**, 190
[41] Bondi H, van der Burg M G J and A. Metzner W K 1962 *Proc. Roy.Soc. A* **269**, 21
[42] Herrera L and Hernández-Pastora J L 2000 *Class. Quantum.Grav.* **17**, 3617
[43] Sorge F, Bini D, de Felice F 2001 *Class. Quantum Grav.***18** 2945
[44] Valiente J 2001 *Class. Quantum. Grav.* **18**, 4311
[45] Herrera L, Santos N O and Carot J 2006 *J. Math.Phys* **47** 052502
[46] Herrera L, Barreto W, Carot J and Di Prisco A 2007 *Class. Quantum.Grav.* **24**, 2645
[47] Bonnor W B 1991 *Phys. Lett. A* **158**, 23
[48] Feynman R P, Leighton R B and Sand M 1964 *Lectures on Physics* **II** (Reading: Addison–Wesley) pp. 27, 28
[49] Herrera L, González G A, Pachón L A and Rueda J A 2006 *Class. Quantum Grav.* **23**, 2395
[50] Bel L 1958 *C. R. Acad. Sci.* 247, 1094
[51] Bel L 1962 *Cah.de Phys.* **16** 59 and re-printed Bel L (2000). *Gen. Rel.Grav.* **32**, 2047
[52] García-Parrado–Gómez-Lobo A 2008 *Class. Quantum Grav.* **25**, 015006
[53] Herrera L, A Di Prisco and Carot J 2007 *Phys. Rev. D* **76**, 044012
[54] Herrera L, Barreto W 2012 *Phys. Rev. D* **86**, 064014
[55] Everitt C F, DeBra D B, Parkinson B W, Turneaure J P, Conklin J W, Heifetz M I, Keiser G M, Silbergleit A S, Holmes T, Kolodziejczak J and Al-Meshari M 2011 *Phys. Rev. Lett.* **106**, 221101
[56] Ciufolini I and Pavlis E 2004 *Nature (London)* **431**, 958
[57] Paolozzi A, Pavlis E C, Ries J C, Koenig R, Matzner R A, Sindoni G and Neumayer H 2009 *Space. Sci. Rev.* **148**, 71
[58] Stedman G E, Schreiber K U and Bilger H R 2003 *Class. Quantum Grav.* **20**, 2527
[59] Stedman G E, Hurst R B and Schreiber K U 2007 *Opt. Commun.* **279**, 124
[60] Scully M O, Zubairy M S and Haugan M P 1981 *Phys. Rev. A* **24**, 2009
[61] Di Virgilio A, Schreiber K U, Gebauer A, Wells J P, Tartaglia A, Belfi J, Beverini N and Ortolan A 2010 *Int. J. Mod. Phys. D* **19**, 2331
[62] Bosi F, Cella G, Di Virgilio A, Ortolan A, Porzio A, Solimeno S, Cerdonio M, Zendri J P, Allegrini M, Belfi J and Beverini N 2011 *Phys. Rev. D* **84**, 122002
[63] Lenef A, Hammond T D, Smith E T, Chapman M S, Rubenstein R A and Pritchard D E 1997 *Phys. Rev. Lett.* **78**, 760
[64] Gustavson T L, Bouyer P and Kasevich M A 1997 *Phys. Rev. Lett.* **78**, 2046
[65] Adler R J, Chen P and Varani E 2012 *Phys. Rev. D* **85**, 025016
[66] Robins N P, Altin P A, Debs J E and Close J D 2012 *Preprint cond-mat.quant-gas:1209.2172v1*
[67] Tasson J D 2012 *Preprint hep-ph:1211.4850v1*
[68] Cambell W C and Hamilton P 2016 *Preprint physics.atom-ph: 1609.00659v1*
[69] Bonnor W B 1965 in *Proceedings of the Meeting on General Relativity*, Florence, ed. G. Barbera (Florence, 1965), Vol.II;
[70] DeWitt B S and Brehme R 1960 *Ann. Phys.(NY)* **9**, 220

[71] Kundt W and Newman E T 1968 *J. Math. Phys.* **9**, 2193
[72] Newman E T and Penrose R 1968 *Proc. R. Soc.*, **A305**, 175
[73] Couch W, Torrence R, Janis A and Newman E T 1968 *J.Math.Phys.* **9**, 484
[74] Bonnor W B 1974 in *Ondes et Radiations Gravitationelles*, (Paris: Editions du Centre National de la Recherche Scientifique)
[75] Blanchet L and Shafer G 1993 *Class.Quantum.Grav.* **10**, 2699
[76] Marchand T, Blanchet L and Faye G 2016 *Class. Quantum. Grav.* **33**, 244003
[77] Sachs R 1962 *Proc. Roy.Soc. A* **270**, 103
[78] Newman E T and Penrose R 1962 *J. Math. Phys.* **3**, 566
[79] Bonnor W B and Rotemberg M 1966 *Proc.R.Soc.* **A289**, 247
[80] Di Prisco A, Herrera L, Jiménez J, Galina V and Ibáñez J 1987 *J. Math. Phys.*, **28**, 2692
[81] Herrera L, Di Prisco A, Ibáñez J and Ospino J 2014 *Phys. Rev. D* **89**, 084034
[82] Ellis G F R and van Ellst H 1998 *Preprint gr-qc: 9812046v4*
[83] Ellis G F R 2009 *Gen. Rel. Grav.* **41**, 581
[84] Ellis G F R , Maartens R and MacCallum M A H 2012 *Relativistic Cosmology* (Cambridge: Cambridge University Press)
[85] Herrera L, Ospino J, Di Prisco A, Fuenmayor E and Troconis O 2009 *Phys. Rev. D* **79**, 064025
[86] Herrera L, Di Prisco A and Ospino J 2012 *Gen. Relativ. Gravit.* **44**, 2645
[87] Herrera L, Di Prisco A, Ospino J and Carot J 2010 *Phys. Rev.D* **82**, 024021
[88] Herrera L, Di Prisco A and Ospino J 2010 *Gen.Relativ. Gravit.* **42**, 1585
[89] Herrera L, Di Prisco A and Ibáñez J 2011 *Phys. Rev. D* **84**, 064036
[90] Herrera L, Di Prisco A and Ibáñez J 2011 *Phys. Rev. D* **84**, 107501
[91] Sharif M and Bashir N 2012 *Gen. Relativ. Gravit.* **44**, 1725
[92] Sharif M and Zaeem Ul Haq Bhatti M 2012 *Mod. Phys. Lett. A* **27**, 1250141
[93] Sharif M and Zaeem Ul Haq Bhatti M 2012 *Gen. Relativ. Gravit.* **44**, 2811
[94] Sharif M and Yousaf Z 2014 *Astr. Space. Sci.* **352**, 321
[95] Sharif M and Yousaf Z 2014 *Astr. Space. Sci.* **354**, 431
[96] Herrera L, Di Prisco A and Ospino J 2014 *Phys. Rev. D* **89**, 127502
[97] Barnes A 1973 *Gen Rel. Gravit* **4**, 105
[98] Barnes A 1984 in *Classical general relativity*. Eds. W.B Bonnor, J.N. Islam and M.A.H. MacCallum (Cambridge: Cambridge University Press)
[99] Glass E N 1975 *J. Math. Phys.* **16**, 2361
[100] Coley A A and McManus, D J 1994 *Class. Quantum Grav.* **11**, 1261
[101] McManus, D J and Coley A A 1994 *Class. Quantum Grav.* **11**, 2045
[102] Di Prisco A, Herrera L, MacCallum M A H and Santos N O 2009 *Phys. Rev. D* **80**, 064031
[103] Herrera L, Di Prisco A, Ospino J and Carot J 2015 *Phys. Rev. D* **91**, 024010
[104] Herrera L, Di Prisco A and Ospino J 2015 *Phys. Rev. D* **91**, 124015
[105] Herrera L, Di Prisco A and Ospino J 2016 *Eur. Phys. J. C* **76**, 603
[106] Herrera L, Di Prisco A, Ospino J and Carot J 2016 *Phys. Rev.D* **94**, 064072
[107] Herrera L, Di Prisco A and Ospino J 2017 *Preprint gr-qc:1702.00983.*
[108] Herrera L, Di Prisco A, Ibáñez J and Ospino J 2013 *Phys. Rev. D* **87**, 024014
[109] Herrera L, Di Prisco A, Ospino J and Carot J (2016). *Int. J. Mod. Phys. D* **25**, 1650036

Power counting renormalization of Hořava gravity at the kinetic conformal point

Jorge Bellorín[12], **Alvaro Restuccia**[12], **Adrián Sotomayor**[3]
[1] Physics Department, Universidad de Antofagasta, Chile
[2] Physics Department, Universidad Simón Bolívar, Caracas, Venezuela
[3] Department of Mathematics, Universidad de Antofagasta, Chile

E-mail: jbellori@gmail.com,arestu@usb.ve,adrian.sotomayor@uantof.cl

Abstract. We show that the Hořava gravity at the kinetic conformal point is power counting renormalizable. The proof is based on the solution, in perturbation theory, of the second class constraints of the formulation.

1. Introduction

Hořava's proposal for a gravity theory assumes an anisotropic behavior of space and time. Consequently, the Lorentz symmetry is manifestly broken in this theory. The benefit would be to obtain a renormalizable theory of gravity at the UV regime. In addition, one may demand that at low energies the relativistic scale should be recovered together with a theory of gravity very close to General Relativity (GR), in order to satisfy the very good agreement of GR to the known experimental data.

The geometrical formulation of the theory is in terms of a foliation of spacelike leaves indexed by a time coordinate. The theory should be invariant under diffeomorphisms on the spacelike leaves (F Diff) together with reparametrization of the time coordinate. Although the use of ADM variables are very adequate to describe the theory, one should have in mind that there is no space-time in Hořava's formulation of gravity. The fields are a spacelike metric denoted g_{ij}, a scalar and a vector under diffeomorphisms of the leaves which transform as densities under time reparametrization, they are denoted N and N^i respectively (the lapse and the shift in the ADM nomenclature).

The theory allows as potential terms in the Hamiltonian higher order spacelike covariant derivatives of the three dimensional Riemann tensor as well as higher order products of the Riemann tensor, which are only allowed to exist in a relativistic theory provided higher order time derivatives terms are also included. In the latest case, generically, there appear ghost fields which ruins the unitary behavior of the theory. Together with the introduction of higher order spacelike covariant derivatives or products of the Riemann tensor in the potential one can include covariant derivatives or products of the vector field $a_i \equiv \frac{\partial_i N}{N}$.

The goal of the Hořava theory of gravity, which was from the beginning to obtain a renormalizable theory of gravity, has not yet been fulfilled. It has been claimed that the theory is power counting renormalizable, but even this point has been argued in terms of analogies with scalar field theories. Gravity theories contain constraints, of the first and second class in Hořava theories, consequently the argument of power counting renormalization has to be proven by solving the constraints and showing that the correct powers of derivatives are involved in the resolution. This point has been recently proven in [1] where a complete argument showing the power counting renormalization of the Hořava gravity at the kinetic conformal point has been obtained. In this work we discuss furthermore several aspects of this proof.

The Hořava theories contain a kinetic term which depends on a dimensiononless coupling constant λ. The so called non-projectable Hořava gravities depend crucially on the value of this coupling constant. For $\lambda \neq \frac{1}{3}$ the theory contains one additional propagating degree of freedom compared to GR. There has been a discussion whether the theory presents problems of strong coupling or not, associated to this propagating mode. For $\lambda = \frac{1}{3}$ the theory contains exactly the same propagating degrees of freedom as GR, the transverse traceless components of the metric on the spacelike leaves [2].

The kinetic term of the theory at $\lambda = \frac{1}{3}$ is invariant under a local conformal transformation. For this reason we have called this theory "Hořava gravity at the kinetic conformal (KC) point". The complete theory is not invariant under that symmetry, only it's kinetic term.

2. The action

The action of the non-projectable Hořava gravity for any value of λ and including the Blass et al. interacting terms is given by [3, 4]

$$S = \int dtd^3x\sqrt{g}N\left(\frac{1}{2\kappa}G^{ijkl}K_{ij}K_{kl} - \mathcal{V}\right), \tag{1}$$

where

$$K_{ij} = \frac{1}{2N}(\dot{g}_{ij} - 2\nabla_{(i}N_{j)})\,, \tag{2}$$

$$G^{ijkl} = \frac{1}{2}\left(g^{ik}g^{jl} + g^{il}g^{jk}\right) - \lambda g^{ij}g^{kl} \tag{3}$$

and λ is a dimensionless constant. Two comments are in order: first, if $z = d = 3$, κ becomes a dimensionless coupling constant [3]. Second, in a relativistic theory, we would have $\lambda = 1$, $z = 1$ and κ would be dimensionful. We do not put a constant in front of the potential $\mathcal{V}$ because we are going to include an independent coupling constant for each one of its terms.

The potential $\mathcal{V}$ can be, in principle, any F Diff scalar made with the spatial metric g_{ij}, the vector

$$a_i = \frac{\partial_i N}{N} \tag{4}$$

and their FDiff-covariant derivatives (curvature tensors and their derivatives for g_{ij}). The potential contains no time derivatives and does not depend on N_i. In particular, the $z = 1$ potential, which is the most relevant one for the large-distance physics, is

$$\mathcal{V}^{(z=1)} = -\beta R - \alpha a_i a^i\,, \tag{5}$$

where β and α are coupling constants.

The kinetic term $\sqrt{g}NG^{ijkl}K_{ij}K_{kl}$ has dimensions -2z where z is the scaling of time. The measure dtd^Dx has dimension $z + D$, hence the overall coupling constant κ must have dimension $-z + D$. It becomes dimensionless for $z = D$. Consequently, the potential should contain all interaction terms compatible with the local symmetry of the theory up to $2D$ derivatives, for $D = 3$ up to six derivatives. The property of κ being dimensionless is crucial in order to show power counting renormalization of the theory.

IOP Conf. Series: Journal of Physics: Conf. Series **831** (2017) 012002 doi:10.1088/1742-6596/831/1/012002

We denote by π^{ij} the momentum conjugated of g_{ij} and by P_N the one of N, whereas we regard the shift vector N_i as a Lagrange multiplier. We study the asymptotically flat case, under which the canonical field variables behave asymptotically as

$$g_{ij} - \delta_{ij} = \mathcal{O}(1/r)\,, \qquad \pi^{ij} = \mathcal{O}(1/r^2)\,, \qquad N - 1 = \mathcal{O}(1/r)\,. \tag{6}$$

The only local constraint associated to gauge symmetries that are homotopic to the identity, and hence of first class, is the momentum constraint $\mathcal{H}^i$,

$$\mathcal{H}^i \equiv -2\nabla_j \pi^{ij} + P_N \partial^i N = 0\,, \tag{7}$$

which generates the purely spatial diffeomorphisms. The second-class constraints are

$$P_N = 0\,, \tag{8}$$
$$\pi \equiv g^{ij}\pi_{ij} = 0\,, \tag{9}$$
$$\mathcal{H} \equiv \frac{2\kappa}{\sqrt{g}}\pi^{ij}\pi_{ij} + \sqrt{g}\,\mathcal{U} = 0\,, \tag{10}$$
$$\mathcal{C} \equiv \frac{3\kappa}{\sqrt{g}}\pi^{ij}\pi_{ij} - \sqrt{g}\,\mathcal{W} = 0\,. \tag{11}$$

$\mathcal{U}$ and $\mathcal{W}$ are derivatives of the potential defined by[1]

$$\mathcal{U} \equiv \frac{1}{\sqrt{g}}\frac{\delta}{\delta N}\int d^3y\sqrt{g}N\mathcal{V} = \mathcal{V} + \frac{1}{N}\sum_{r=1}(-1)^r \nabla_{i_1\cdots i_r}\left(N\frac{\partial \mathcal{V}}{\partial(\nabla_{i_r\cdots i_2}a_{i_1})}\right), \tag{12}$$

$$\mathcal{W} \equiv g_{ij}\mathcal{W}^{ij}\,, \qquad \mathcal{W}^{ij} \equiv \frac{1}{\sqrt{g}N}\frac{\delta}{\delta g_{ij}}\int d^3y\sqrt{g}N\mathcal{V}\,. \tag{13}$$

$\nabla_{ij\cdots k}$ stands for $\nabla_i\nabla_j\cdots\nabla_k$. Adopting the nomenclature of GR, $\mathcal{H}^i = 0$ is called the momentum constraint and $\mathcal{H} = 0$ the Hamiltonian constraint.

The $\pi = 0$ constraint is the primary constraint that emerges when the theory is formulated at the KC point. Indeed, the conjugated momentum π^{ij} obeys the general relation

$$\frac{\pi^{ij}}{\sqrt{g}} = \frac{1}{2\kappa}G^{ijkl}K_{kl}\,. \tag{14}$$

At $\lambda = 1/3$ the hypermatrix G^{ijkl} becomes degenerated, $g_{ij}G^{ijkl} = 0$, which leads directly to the $\pi = 0$ constraint. As a consequence, the secondary constraint $\mathcal{C} = 0$ emerges when the preservation in time of $\pi = 0$ is demanded. Thus, π and $\mathcal{C}$ are the two second-class constraints that emerge at the KC point. Unlike GR, in the nonprojectable Hořava theory the Hamiltonian constraint $\mathcal{H}$ is of second-class behavior, which is associated to the fact that it lacks its role as generator of gauge symmetry. Finally, the $P_N = 0$ constraint must be added since in this theory (with $\lambda = 1/3$ or not) we are forced to included the lapse function N as part of the canonical variables.

In addition, the boundary term corresponding to the ADM energy [5],

$$E_{\text{ADM}} \equiv \oint d\Sigma_i(\partial_j g_{ij} - \partial_i g_{jj})\,, \tag{15}$$

[1] We have modified the original definition of $\mathcal{C}$ given in Ref. [2] by dividing it by N.

must be incorporated because it is needed for the differentiability of the Hamiltonian under the most general asymptotic variations compatible with asymptotic flatness [6, 7].

By incorporating the constraints P_N and π, we finally cast the classical Hamiltonian in the form

$$H = \int d^3x \left(\frac{2\kappa N}{\sqrt{g}} \pi^{ij}\pi_{ij} + \sqrt{g} N \mathcal{V} + N_i \mathcal{H}^i + \sigma P_N + \mu\pi \right) + \beta E_{\text{ADM}} , \tag{16}$$

where N_i, σ and μ are Lagrange multipliers. This classical Hamiltonian is subject to the constraints (10) and (11), which have not been added with Lagrange multipliers.

In the counting of the independent degrees of freedom we have 14 non-reduced canonical variables in the set $\{(g_{ij}, \pi^{ij}), (N, P_N)\}$, three components of the first-class constraint $\mathcal{H}^i$ and four second-class constraints in the set $\{P_N, \pi, \mathcal{H}, \mathcal{C}\}$. The number of independent degrees of freedom is given by

$$(14 \text{ can. var.}) - [2 \times (3 \text{ first-cls. c.}) + (4 \text{ second-cls. c.})] = 4 \text{ indep. can. var.} \tag{17}$$

Thus, there are two even physical modes in the theory; that is, two modes that propagate themselves with a complete pair of canonical variables. This is the same number of degrees of freedom of GR; there are no extra modes in this theory. This property naturally raises the question whether the dynamics of this theory is able to reproduce the dynamics of GR for suitable large distances, i. e., at least in a perturbative regime for both theories. This was analyzed for the perturbatively linearized theory in Ref. [2].

3. The solution of the constraints at the linearized level

We present in this section the perturbative approach around Minkowski space-time and the solution of the constraints at the linearizad level.

The complete set of non-equivalent terms which contribute to the propagator of the theory is [8],

$$-\mathcal{V}^{(z=1)} = \beta R + \alpha a_i a^i , \tag{18}$$

$$-\mathcal{V}^{(z=2)} = \alpha_1 R \nabla_i a^i + \alpha_2 \nabla_i a_j \nabla^i a^j + \beta_1 R_{ij} R^{ij} + \beta_2 R^2 , \tag{19}$$

$$-\mathcal{V}^{(z=3)} = \alpha_3 \nabla^2 R \nabla_i a^i + \alpha_4 \nabla^2 a_i \nabla^2 a^i + \beta_3 \nabla_i R_{jk} \nabla^i R^{jk} + \beta_4 \nabla_i R \nabla^i R , \tag{20}$$

where $\nabla^2 \equiv \nabla_i \nabla^i$ and all the alphas and betas are coupling constants.

The perturbations around Minkowski spacetime are defined by introducing the variables h_{ij}, p_{ij} and n in the following way

$$g_{ij} = \delta_{ij} + \epsilon h_{ij} , \qquad \pi^{ij} = \epsilon p_{ij} , \qquad N = 1 + \epsilon n . \tag{21}$$

We use the orthogonal transverse/longitudinal decomposition

$$h_{ij} = h^{TT}_{ij} + \frac{1}{2}(\delta_{ij} - \partial_{ij}\partial^{-2}) h^T + \partial_{(i} h^L_{j)} , \tag{22}$$

where $\partial_{ij\cdots k}$ stands for $\partial_i \partial_j \cdots \partial_k$, $\partial^2 = \partial_i \partial_i$ and $\partial^{-2} = 1/\partial^2$. h^{TT}_{ij} is subject to $\partial_i h^{TT}_{ij} = h^{TT}_{ii} = 0$. We make an analogous decomposition on p_{ij}. We impose the transverse gauge,

$$\partial_i h_{ij} = 0 , \tag{23}$$

under which all the longitudinal sector of the metric is eliminated.

We study the constraints (7 - 11) of the theory at linear order in perturbations in terms of the potential defined in (18 - 20). The momentum constraint (7), simplified by using $P_N = 0$ explicitly, eliminates the longitudinal sector of p_{ij},

$$\partial_i p_{ij} = 0\,, \tag{24}$$

whereas the $\pi = 0$ constraint dictates that p_{ij} is traceless, hence $p^T = 0$. So far we are left with the set $\{h_{ij}^{TT}, p_{ij}^{TT}, h^T, n\}$ as the set of remaining canonical variables.

Now we consider the $\mathcal{H}$ and $\mathcal{C}$ constraints. To present the results in a compact form, we introduce the vector ϕ of scalars and the functional matrix $\mathbb{M}$ in the way

$$\phi = \begin{pmatrix} h^T \\ n \end{pmatrix}, \qquad \mathbb{M} = \begin{pmatrix} \mathbb{D}_1 & \mathbb{D}_2 \\ \mathbb{D}_2 & \mathbb{D}_3 \end{pmatrix}, \tag{25}$$

where

$$\begin{aligned} \mathbb{D}_1 &\equiv \frac{1}{8}\left((3\beta_3 + 8\beta_4)\partial^6 - (3\beta_1 + 8\beta_2)\partial^4 + \beta\partial^2\right), \\ \mathbb{D}_2 &\equiv \frac{1}{2}\left(\alpha_3\partial^6 + \alpha_1\partial^4 + \beta\partial^2\right), \qquad \mathbb{D}_3 \equiv \alpha_4\partial^6 - \alpha_2\partial^4 + \alpha\partial^2\,. \end{aligned} \tag{26}$$

Thus, with the potential given in (18 - 20), the $\mathcal{H}$ and $\mathcal{C}$ constraints at linear order become

$$\mathbb{M}\phi = 0\,, \tag{27}$$

where the first row of this vectorial equation represents the $\mathcal{C}$ constraint and the second row the $\mathcal{H}$ constraint.

To solve the constraints (27) we start by decoupling them; that is, we want two separate equations in which h^T and n are not mixed. To this end we multiply Eq. (27) with

$$\begin{pmatrix} \mathbb{D}_3 & -\mathbb{D}_2 \\ -\mathbb{D}_2 & \mathbb{D}_1 \end{pmatrix} \tag{28}$$

from the left and get a diagonal matrix acting on ϕ, which we write as

$$\mathbb{L}\phi = 0\,, \qquad \mathbb{L} \equiv \left(\mathbb{D}_1\mathbb{D}_3 - \mathbb{D}_2^2\right)\mathbb{I}\,. \tag{29}$$

Equation (29) represents two decoupled equations for h^T and n and, moreover, the equations are the same (with the same boundary conditions).

Given the values of all the coupling constants, the generic case is when the operator $\mathbb{L}$ is a sixth-order polynomial on ∂^2. We can always factorize it; in particular, we may write it as

$$\mathbb{L} = K(\partial^2 - z_1)P^{(5)}(\partial^2)\,, \tag{30}$$

where $P^{(5)}(u)$ is a fifth-order polynomial on u, z_1 stands for any one of the roots of $\mathbb{L}$, where $K = (1/8)\left(\alpha_4(3\beta_3 + 8\beta_4) - 2\alpha_3^2\right)$. If $K \neq 0$ by combining (30) with (29) we can write the constraints in the form

$$\partial^2 P^{(5)}(\partial^2)\phi = z_1 P^{(5)}(\partial^2)\phi\,. \tag{31}$$

The decoupled equation (31) implies that $P^{(5)}(\partial^2)\phi$ is an eigenfunction of the Laplacian ∂^2. Since we are studying the asymptotically flat case, the spatial domain of the problem is the whole $\mathbb{R}^3$ and the boundary condition is that ϕ and its derivatives are zero at spatial infinity.

Consider the operator $\partial^2 - z_1$, with $z_1 \in \mathbb{C}$, acting on the space of functions ψ whose domain is the whole $\mathbb{R}^3$ and that go asymptotically to zero (see (6)). Thus, Eq. (31) can be cast as

$$(\partial^2 - z_1)\psi = 0\,. \tag{32}$$

In the space of functions ψ, ∂^2 has a continuum spectrum valued in $(-\infty, 0]$; it has no eigenvalues. With the prescribed asymptotic behavior the inverse $(\partial^2 - z_1)^{-1}$ exists for any value of z_1, but it behaves in different ways depending on whether z_1 belongs to the spectrum or not. If $z_1 \notin (-\infty, 0]$ the inverse $(\partial^2 - z_1)^{-1}$ is a bounded operator. In this case Eq. (32) automatically implies $\psi = 0$. If $z_1 \in (-\infty, 0]$, $(\partial^2 - z_1)^{-1}$ still exists but it is an unbounded operator. However, the right-hand side of Eq. (32) is zero; $(\partial^2 - z_1)^{-1}$ acting on it gives zero anyway. Therefore, for any value of z_1, Eq. (32) has the function $\psi = 0$ as its only solution satisfying the prescribed asymptotic behavior. We then have

$$P^{(5)}(\partial^2)\phi = 0. \tag{33}$$

Coming back to Eq. (33), it turns out that it becomes another eigenfunction problem for the Laplacian since its left-hand side is another polynomial on ∂^2, such that we may factorize it again,

$$(\partial^2 - z_2)\, P^{(4)}(\partial^2)\phi = 0\,. \tag{34}$$

Since the same arguments hold to solve this equation, we have $P^{(4)}(\partial^2)\phi = 0$ as the unique solution. We may proceed iteratively continuing with this last equation to finally show that the linear-order versions for the variables h^T and n are equal to zero.

We remark that it is the non-compactness of the domain and the prescribed asymptotic conditions of the problem posed in (31) that force the everywhere-vanishing function to be the unique eigenfunction.

If $\mathbb{L}$ is a lower order polynomial ($K = 0$), an analogous eigenfunction problem for the Laplacian arises since we may factorize the given polynomial. By applying the same reasoning of above, we eventually arrive at the same zero solution. Therefore, we conclude that the unique solution of the linearized $\mathcal{H}$ and $\mathcal{C}$ constraints, which are expressed in (27), is

$$h^T = n = 0\,. \tag{35}$$

There remains a condition in the space of parameters: we require that the whole operator $\mathbb{L}$ is not completely zero since otherwise the number of constraints effectively reduces and additional modes appear. In addition, we know that the perturbatively linearized version of the purely $z = 1$ theory is equivalent to perturbatively linearized GR [2]. To combine these two facts, we require that the fourth-order coefficient of $\mathbb{L}$, associated to the $z = 1$ operators of the theory, is nonzero,

$$\beta(2\beta - \alpha) \neq 0\,. \tag{36}$$

Following the analysis of the linearized theory, it can be shown that the Lagrange multipliers μ and σ, and n_i the perturbation of $N_i = 0$, are zero at the linearized level.

We finally have that, when all the constraints have been solved and the gauge has been fixed at linear order, there remains the pair $\{h_{ij}^{TT}, p_{ij}^{TT}\}$ as the set of free canonical variables. This confirms rigorously the number of two propagating degrees of freedom that the generic and nonperturbative Hamiltonian analysis anticipated.

4. The Hamiltonian at quadratic order in perturbations

Once we know the solutions of all the constraints in the transverse gauge, we may compute the reduced canonical Hamiltonian of the linearized theory.

It is given by

$$H_{\text{RED}} = \int d^3x \left(2\kappa p_{ij}^{TT} p_{ij}^{TT} + \frac{1}{4} h_{ij}^{TT} \mathbb{V} h_{ij}^{TT} \right), \tag{37}$$

where

$$\mathbb{V} = -\beta\partial^2 - \beta_1\partial^4 + \beta_3\partial^6\,. \tag{38}$$

The large-distance dynamics of the perturbatively linearized theory can be obtained from the reduced Hamiltonian (37) by neglecting the higher order derivatives against the lowest order one. By doing so we obtain the effective Hamiltonian for the tensorial modes

$$H^{\rm eff}_{\rm RED} = \int d^3x \left(2\kappa p^{TT}_{ij} p^{TT}_{ij} - \frac{\beta}{4} h^{TT}_{ij} \partial^2 h^{TT}_{ij} \right) . \tag{39}$$

This is equivalent to taking only the $z = 1$ potential (5) and then linearizing it [2]. Thus, the perturbatively linearized version of the large-distance effective action is physically equivalent to linearized GR. Here one of the key features is the vanishing of the variables h^T and n at linear order in perturbations. The evolution equations arising from (39) constitute the wave equation for h^{TT}_{ij}, thus the perturbative large-distance theory around Minkowski spacetime propagates gravitational waves exactly as linearized GR does. However, the nonperturbative dynamics of both theories are different, even considering only the $z = 1$ order in the side of the Hořava theory, since the nonperturbative field equations are different.

The requirement of positivity of the reduced Hamiltonian imposes constraints on the coupling constants β, β_1 and β_3 (we assume that κ is positive). We require that $\mathbb{V} \geq 0$.

The resulting restrictions on the coupling constants needed for the continuity in the number of degrees of freedom, weakest regime approaching to GR, positivity and $z = 3$ behavior of the Hamiltonian are

$$\alpha \neq 2\beta \,, \qquad \beta > 0 \,, \qquad \beta_3 < 0 \,, \qquad \beta_1 \leq 2\sqrt{\beta|\beta_3|} \,. \tag{40}$$

5. The propagator of the physical modes

From the expression (37), we obtain the full propagator of the physical modes, it is given by

$$\langle h^{TT}_{ij} h^{TT}_{kl} \rangle = \frac{P^{TT}_{ijkl}}{\omega^2/2\kappa - \beta \vec{k}^2 + \beta_1 \vec{k}^4 + \beta_3 \vec{k}^6} \,, \tag{41}$$

where

$$P^{TT}_{ijkl} \equiv \frac{1}{\sqrt{2}} \left(\theta_{ik}\theta_{jl} + \theta_{il}\theta_{jk} - \theta_{ij}\theta_{kl} \right) , \qquad \theta_{ij} \equiv \delta_{ij} - \frac{k_i k_j}{\vec{k}^2} \,. \tag{42}$$

Notice that only some terms of the potential (18 - 20) contribute to the propagator of the physical modes. The independent propagator (41) of this theory behaves just as was the aim in the original formulation of Hořava for having a renormalizable and unitary theory of quantum gravity [3]: for high ω and $\vec{k}$ it is dominated by the $z = 3$ mode $(\omega^2/2\kappa + \beta_3 \vec{k}^6)^{-1}$ and there are no more independent propagators other than (41).

Besides, at low energies the dominant term is $\left(\frac{w^2}{2k} - \beta \vec{k}^2 \right)^{-1}$. That means, for appropiate values of the couplings we recover the propagator of the physical modes of GR.

6. Power counting renormalization

We will analyze the theory at the *UV* regime. This requires to go beyond the linear order in the solution of the constraints.

We use the notation in (25). We have for the next order in perturbations

$$\epsilon \mathbb{M} \phi = \epsilon^2 \mathbb{T}(p^{TT}_{ij}, h^{TT}_{ij})$$

where $\mathbb{T}$ is a 2×2 matrix with entries quadratic in p^{TT}_{ij} and quadratic in h^{TT}_{ij} and in its spacelike derivatives. Because of the dimensions there are no derivatives of p^{TT}_{ij}. The solution at linear order for h^T and n is zero, hence $\mathbb{T}$ depends only on p^{TT}_{ij} and h^{TT}_{ij}.

The relevant point is that the highest order in derivatives ∂^6 appears on the left hand member of the equation. It is given by a matrix which can be identified from $\mathbb{M}$, its components are dimensionless and its determinant is K. We must assume $K \neq 0$. Then, when solving for ϕ the solution has a maximal order on spacelike derivatives on the denominator. This ensures that the solution for h^T and n, which has to be inserted on the interacting vertices, do not contribute with positive powers of momenta.

For higher order solutions for h^T and n in the iterative process, we have the same matrix $\mathbb{M}$ in the left hand of the constraints and a similar argument follows. At each iterative step the solution for h^T and n do not contribute with positive powers of momenta. It is crucial in the argument the assumption that the dimensionless $K = (1/8)\left(\alpha_4(3\beta_3 + 8\beta_4) - 2\alpha_3^2\right) \neq 0$, otherwise the argument of power counting renormalization fails. This point shows that arguments based in analogies with scalar fields theories are only heuristic ones which are not valid in general.

We may now discuss the power-counting renormalizability guided by the superficial degree of divergence of general 1PI diagrams over the reduced phase space. For this computation we follow Refs. [9, 10]. Further developments on the renormalization of Lorentz-violating theories, in particular, studies on the behavior of the subdivergences, were made in Refs. [11]. From the propagator (41) we deduce that if Λ is an UV cutoff for the momenta, then Λ^z is the cutoff for the energy (up to some constants of proportionality that are irrelevant for our purposes), with $z = 3$. Therefore, for each loop in the UV regime we have the contribution

$$\int d\omega d^d k \rightarrow \Lambda^{d+z}, \tag{43}$$

while for each propagator

$$I = \Lambda^{2z}. \tag{44}$$

In any vertex we can have at most a contribution of $2z$ powers of loop momenta coming from the vertex itself (for vertices that are of $2z$ order in spatial derivatives). If in a 1PI Feynman diagram L is the number of loops, I is the number of internal lines and V is the number of vertices, its superficial degree of divergence D is bounded by

$$\begin{aligned} D &\leq (d+z)L + 2z(V - I) &(45)\\ &= (d-z)L + 2z(L + V - I). &(46) \end{aligned}$$

Now the identity $L - 1 = I - V$ for graphs is used and in addition in this theory we have $z = d$. Therefore, the superficial degree of divergence is bounded by

$$D \leq 2z. \tag{47}$$

This is the bound (8) of Ref. [10], where Lorentz-violating theories with interactions depending on spatial derivatives were considered. This degree of divergence coincides with the highest order operators already included in the bare action (once we extend our potential to include all the $z \leq 3$ terms, not only the operators that contribute to the quadratic action). This leads to the conclusion that the theory is power-counting renormalizable. Unitarity and the criterion of power-counting renormalizability are safe in this theory.

7. Conclusions

We showed, under the assumption $K = (1/8)\left(\alpha_4(3\beta_3 + 8\beta_4) - 2\alpha_3^2\right) \neq 0$ on the dimensionless coupling constants, that Hořava gravity at the kinetic conformal point propagates the same physical degrees of freedom as General Relativity and it is power counting renormalizable. The results are based on previous works [1, 2, 12].

Acknowledgments

A. R. and A. S. are partially supported by Project Fondecyt 1161103, Chile.

References

[1] J. Bellorín an Alvaro Restuccia, *Quantization of the Horava theory at the kinetic-conformal point*, Phys. Rev. D **94** (2016) 064041 [arXiv:1606.02606 [hep-th]].

[2] J. Bellorín, A. Restuccia and A. Sotomayor, *A consistent Hořava gravity without extra modes and equivalent to general relativity at the linearized level*, Phys. Rev. D **87** (2013) 084020 [arXiv:1302.1357 [hep-th]].

[3] P. Hořava, *Quantum Gravity at a Lifshitz Point*, Phys. Rev. D **79** (2009) 084008 [arXiv:0901.3775 [hep-th]].

[4] D. Blas, O. Pujolàs and S. Sibiryakov, *Consistent Extension Of Hořava Gravity*, Phys. Rev. Lett. **104** (2010) 181302 [arXiv:0909.3525 [hep-th]].

[5] R. L. Arnowitt, S. Deser and C. W. Misner, *The Dynamics of general relativity*, Gen. Rel. Grav. **40** (2008) 1997 [arXiv:gr-qc/0405109].

[6] T. Regge and C. Teitelboim, *Role of Surface Integrals in the Hamiltonian Formulation of General Relativity*, Annals Phys. **88** (1974) 286.

[7] S. W. Hawking and G. T. Horowitz, *The Gravitational Hamiltonian, action, entropy and surface terms*, Class. Quant. Grav. **13** (1996) 1487 [arXiv:gr-qc/9501014 [gr-qc]].

[8] M. Colombo, A. E. Gümrükçuoğlu and T. P. Sotiriou, *Hořava gravity with mixed derivative terms*, Phys. Rev. D **91** (2015) 4, 044021 [arXiv:1410.6360 [hep-th]].

[9] M. Visser, *Lorentz symmetry breaking as a quantum field theory regulator*, Phys. Rev. D **80** (2009) 025011 [arXiv:0902.0590 [hep-th]].

[10] M. Visser, *Power-counting renormalizability of generalized Horava gravity*, arXiv:0912.4757 [hep-th].

[11] D. Anselmi and M. Halat, *Renormalization of Lorentz violating theories*, Phys. Rev. D **76** (2007) 125011 [arXiv:0707.2480 [hep-th]]. D. Anselmi, *Weighted scale invariant quantum field theories*, JHEP **0802** (2008) 051 [arXiv:0801.1216 [hep-th]]. D. Anselmi, *Weighted power counting and Lorentz violating gauge theories. I. General properties*, Annals Phys. **324** (2009) 874 [arXiv:0808.3470 [hep-th]]. D. Anselmi, *Weighted power counting and Lorentz violating gauge theories. II. Classification*, Annals Phys. **324** (2009) 1058 [arXiv:0808.3474 [hep-th]]. D. Anselmi, *Weighted power counting, neutrino masses and Lorentz violating extensions of the Standard Model*, Phys. Rev. D **79** (2009) 025017 [arXiv:0808.3475 [hep-ph]].

[12] J. Bellorín, A. Restuccia and A. Sotomayor, *Non-perturbative analysis of the constraints and the positivity of the energy of the complete Hořava theory*, Phys. Rev. D **85** (2012) 124060 [arXiv:1205.2284 [hep-th]].

Features in single field slow-roll inflation

Alexander Gallego Cadavid
Instituto de Física, Universidad de Antioquia, A.A.1226, Medellín, Colombia
E-mail: alexander.gallego@udea.edu.co

Abstract. We compare the effects of local features (LF) and branch features (BF) of the inflaton potential on the spectrum of primordial perturbations. We show that LF affect the spectrum in a narrow range of scales while BF produce a step between large and small scales with respect to the featureless spectrum. We comment on the possibility of distinguishing between these two types of feature models from the analysis of the Cosmic Microwave Background (CMB) radiation data.

We also show that there exists a quantitative similarity between the primordial spectra predicted by two of the BF potentials considered. This could lead to a degeneracy of their predicted CMB temperature spectra which could make difficult to discriminate between the models from a CMB analysis. We comment on the possibility that the degeneracy can be broken when higher order terms in the perturbations are considered. In this sense non-Gaussianity may play an important role in discerning between different inflationary models which predict similar spectra.

1. Introduction

There is a significant set of observational data from experiments like the Wilkinson Microwave Anisotropy Probe (WMAP) [1–4], the Planck Satellite [5–9], the Sloan Digital Sky Survey (SDSS) [10], and many other ground-based and sub-orbital experiments [11]. The surveys reveal that the observable Universe is homogeneous and isotropic at large scales.

The Cosmic Microwave Background (CMB) radiation is formed by the relic photons coming from the Last Scattering Surface and emitted when the Universe was only about 380, 000 years old [4–6, 8, 12]. Due to the expansion of the Universe, the temperature decreased and protons and electrons combined to form neutral light atoms (recombination) while photons started to travel freely through space (decoupling). Although this radiation is extremely isotropic, it is still possible to observe small fluctuations in the temperature map of the CMB radiation of the order $\Delta T/T \sim 10^{-5}$ [4–6, 8, 12]. Since at the recombination epoch the Universe was very young, a detailed measurement and subsequent study of this radiation can give us valuable information about the physics of the early Universe.

The extraordinary advances in observable cosmology in the last decades give us evidence that the observed structures originated from seed fluctuations in the very early Universe. Inflationary cosmology is one of the simplest frameworks able to explain the origin of these primordial fluctuations and to provide a good fit to the data [7, 9, 13–16]. It is currently one of the most active research fields in describing the evolution of the very early Universe. Cosmic inflation corresponds to an epoch of accelerated expansion of the Universe. In the simplest models, inflation is driven by a single canonical scalar field, called the *inflaton*, and slowly rolling a smooth potential [7, 9]. During inflation the quantum fluctuations of the scalar field got stretched to cosmological scales which can later be seen as inhomogeneities and anisotropies in the Large Scale Structures (LSS) and the CMB radiation [7,9].

The latest observations have led us to the Standard Model of cosmology, the so called ΛCDM model [4–9]. According to this model the Universe is spatially flat, with *nearly* Gaussian primordial perturbations with a spectrum usually taken to be a *featureless* (smooth) primordial power spectrum (PPS), described by an *almost* scale-invariant power-law [4–6, 8].

However, there exist several well-motivated theoretical inflationary models which predict a non-standard PPS with features [17–27]. Indeed, we already have indications of primordial features in all releases of WMAP [1–4,28] and Planck 2013 and 2015 temperature data [8,9,28], and we have knowledge about the location and type of the features from reconstructions of the PPS [29–32].

Nowadays, primordial features is one of the most exciting extensions of the ΛCDM model [20–24, 28, 30, 33–43]. They can provide a wealth of information about the primordial Universe, ranging from discrimination between inflationary models, distinguishing inflation from alternative scenarios, and even to new particle detection [20–24,28,30,33–36,39]. Moreover it has been shown that features in the PPS can actually improve the fit to Planck CMB temperature and polarization data [19, 28, 32, 38, 44–50].

In this paper we are interested in an important class of models that predict primordial scale-dependent oscillatory features in the PPS. In particular, we study and compare the predictions on the PPS of two different kind of features of the inflaton potential.

The paper is organized as follows. In Sec. 2 we introduce the models and describe their different characteristics. In Sec. 3 we calculate numerically the PPS for these models and discuss their different predictions. We conclude in Sec. 4.

2. The inflationary models

We consider inflationary models with a single scalar field, minimally coupled to gravity, and standard kinetic term according to the action

$$S = \int d^4x\sqrt{-g}\left[\frac{1}{2}M_{Pl}^2 R - \frac{1}{2}g^{\mu\nu}\partial_\mu\phi\partial_\nu\phi - V(\phi)\right], \tag{1}$$

where $g_{\mu\nu}$ is the Friedmann-Lemaitre-Robertson-Walker (FLRW) metric in a flat Universe, $M_{Pl} = (8\pi G)^{-1/2}$ the reduced Planck mass, R the Ricci scalar, and V is the potential energy of the inflaton ϕ. The variation of the action with respect to the metric and the scalar field gives the Friedmann equation and the inflaton equation of motion respectively

$$H^2 \equiv \left(\frac{\dot{a}}{a}\right)^2 = \frac{1}{3M_{Pl}^2}\left(\frac{1}{2}\dot{\phi}^2 + V(\phi)\right), \tag{2}$$

$$\ddot{\phi} + 3H\dot{\phi} + \partial_\phi V = 0, \tag{3}$$

where H is the Hubble parameter, a the scale factor, and we denote the derivatives with respect to time and scalar field with dots and ∂_ϕ, respectively.

For the slow-roll parameters we use the following definitions

$$\epsilon \equiv -\frac{\dot{H}}{H^2} \quad , \quad \eta \equiv \frac{\dot{\epsilon}}{\epsilon H}. \tag{4}$$

To study the effects of features on the spectrum of primordial perturbations we consider three different inflationary models defined by the potentials [19, 35, 51, 52]

$$V_1(\phi) = V_0(\phi) + V_{LF}(\phi), \tag{5}$$

$$V_2(\phi) = V_0(\phi) + V_{BF2}(\phi), \tag{6}$$

$$V_3(\phi) = V_0(\phi)V_{BF3}(\phi), \tag{7}$$

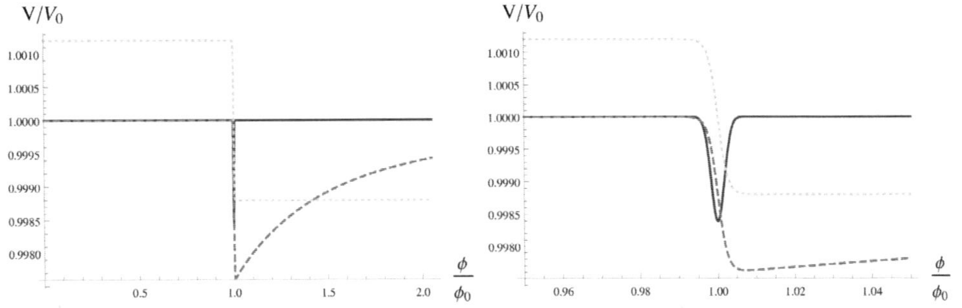

Figure 1. The evolution of $V_i/V_0, (i = 1,2,3)$ as a function of the field is plotted for the potentials in Eqs. (5-7). In all cases we use $\gamma = 1.66 \times 10^{-11}, \sigma = 0.04$, and we set the scale of the feature at $k_0 = 1.5 \times 10^{-3}\text{Mpc}^{-1}$. For V_1, V_2, and V_3 we use $\lambda = -7 \times 10^{-12}$ (blue), $\lambda = -5.25 \times 10^{-12}$ (red-dashed), and $\lambda = -1.2 \times 10^{-3}$ (green-dotted), respectively. On the left panel we can see that the potentials V_2 and V_3 are indeed different from each other for all values of the field. On the right panel we plot a zoom of the evolution around the feature value ϕ_0.

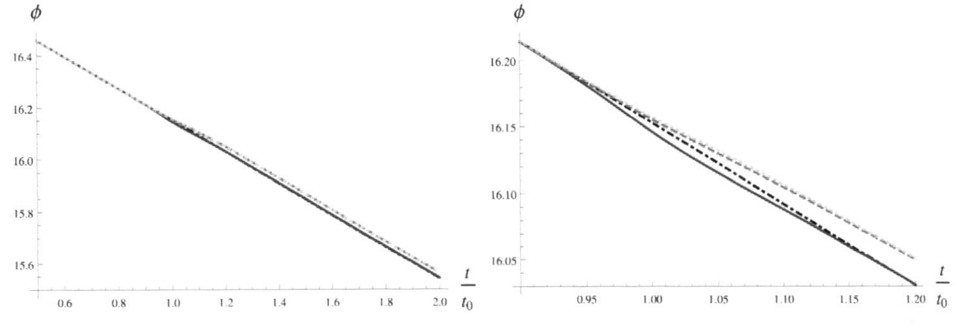

Figure 2. The numerically computed inflaton evolution is plotted for the potentials in Eqs. (5-7). In all cases we use $\gamma = 1.66 \times 10^{-11}, \sigma = 0.04$, and we set the scale of the feature at $k_0 = 1.5 \times 10^{-3}\text{Mpc}^{-1}$. For V_1, V_2, and V_3 we use $\lambda = -7 \times 10^{-12}$ (blue), $\lambda = -5.25 \times 10^{-12}$ (red-dashed), and $\lambda = -1.2 \times 10^{-3}$ (green-dotted), respectively. The dashed-dotted black lines correspond to the featureless case. On the right panel we plot a zoom of the evolution around ϕ_0.

where V_0 is the chaotic inflaton [7,9] potential that we use as a toy model to show qualitatively the general type of effects produced by the features

$$V_0(\phi) = \gamma\phi^2 , \tag{8}$$

and

$$V_{LF}(\phi) = \lambda e^{-(\frac{\phi-\phi_0}{\sigma})^2} , \tag{9}$$

$$V_{BF2}(\phi) = \lambda\left(1 + \tanh\left(\frac{\phi - \phi_0}{\sigma}\right)\right) , \tag{10}$$

$$V_{BF3}(\phi) = 1 + \lambda \tanh\left(\frac{\phi - \phi_0}{\sigma}\right) . \tag{11}$$

In general the parameters λ and σ are related to the amplitude and width of the feature respectively [19, 33–35, 51, 52]. While the parameter $\phi_0 \equiv \phi(t_0)$, where t_0 is the feature

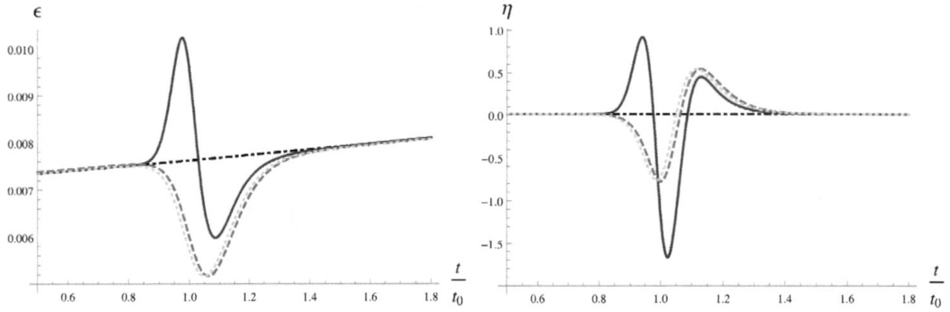

Figure 3. The numerically computed slow-roll parameters are plotted for the potentials in Eqs. (5-7). In all cases we use $\gamma = 1.66 \times 10^{-11}, \sigma = 0.04$, and we set the scale of the feature at $k_0 = 1.5 \times 10^{-3}\text{Mpc}^{-1}$. For V_1, V_2, and V_3 we use $\lambda = -7 \times 10^{-12}$ (blue), $\lambda = -5.25 \times 10^{-12}$ (red-dashed), and $\lambda = -1.2 \times 10^{-3}$ (green-dotted), respectively. The dashed-dotted black lines correspond to the featureless case.

time, is related to the scale of the feature $k_0 \equiv -1/\tau_0$, where τ is the conformal time $d\tau \equiv dt/a$ [19,33–35,51,52]. From now on we adopt a system of units in which $c = \hbar = M_{Pl} = 1$.

The modifications of the slow-roll potential V_1 correspond to local features (LF), first studied in Ref. [35], while the ones in V_2 and V_3 are called branch features (BF) [15,18,19,33–35,51–53]. The BF differ from the LF because their definitions involve step functions or their smoothed versions, which divide the potential in separate branches [35]. For these models the potential is not only modified around the feature ϕ_0, but at any other field value after or before ϕ_0 [35], as shown in figure 1. There it can be seen that the LF potential deviates from the featureless one only in a limited range of the field values around ϕ_0 while the BF potentials differ from the featureless case: before the feature for potential V_2 and before and after the feature for V_3. In figure 2 we show the effects of the features on the inflaton evolution with respect to time. We see again that the BF diverge notoriously from the featureless behavior while the deviation of the LF is only around the feature time t_0. In figure 3 we plot the effects of the features on the slow-roll parameters with respect to time.

Another interesting characteristic of potentials V_2 and V_3 is that, with the parameters chosen, they lead to a similar evolution of the inflaton and slow-roll parameters, as can be seen in figures 2 and 3. The importance of this behavior is that it could lead to a degeneracy of the primordial or CMB spectra which would make harder to distinguish between these different models.

3. Primordial spectrum of curvature perturbations

The study of primordial scalar perturbations is attained by expanding perturbatively the action with respect to the background FLRW solution [54]. In the comoving gauge, where the scalar field fluctuations vanish, $\delta\phi = 0$, the second order action for the scalar perturbations is [54]

$$S_2 = \int dt d^3x \left[a^3 \epsilon \dot{\mathcal{R}}_c^2 - a\epsilon(\partial \mathcal{R}_c)^2\right], \tag{12}$$

where $\mathcal{R}_c$ is the curvature perturbation on comoving slices. The Euler-Lagrange equations for this action give

$$\frac{\partial}{\partial t}\left(a^3 \epsilon \frac{\partial \mathcal{R}_c}{\partial t}\right) - a\epsilon\delta^{ij}\frac{\partial^2 \mathcal{R}_c}{\partial x^i \partial x^j} = 0. \tag{13}$$

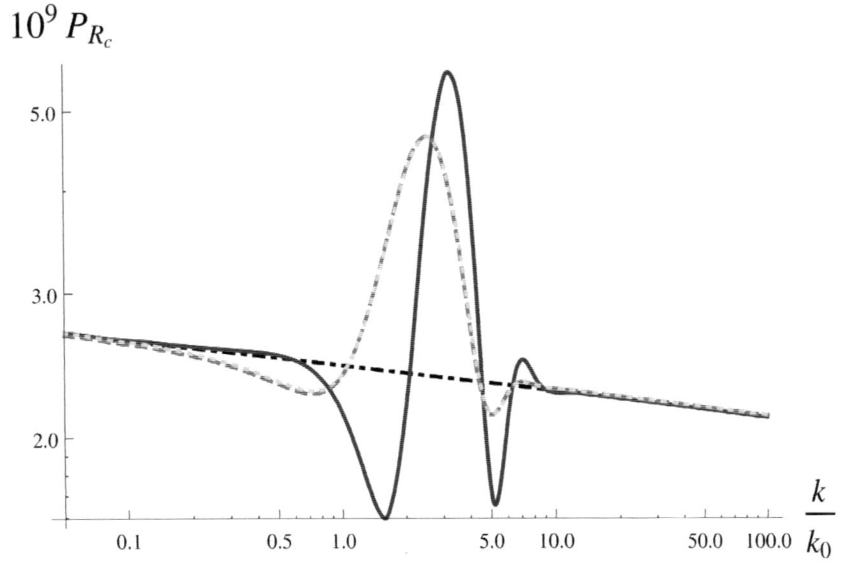

Figure 4. The numerically computed power spectrum of primordial curvature perturbations $P_{\mathcal{R}_c}$ is plotted for the potentials in Eqs. (5-7). In all cases we use $\gamma = 1.66\times10^{-11}, \sigma = 0.04$, and we set the scale of the feature at $k_0 = 1.5\times10^{-3}\text{Mpc}^{-1}$. For V_1, V_2, and V_3 we use $\lambda = -7\times10^{-12}$ (blue), $\lambda = -5.25\times10^{-12}$ (red-dashed), and $\lambda = -1.2\times10^{-3}$ (green-dotted), respectively. The dashed-dotted black lines correspond to the featureless spectrum.

Then taking the Fourier transform we obtain the equation of motion for primordial curvature perturbation modes

$$\mathcal{R}_c''(k) + 2\frac{z'}{z}\mathcal{R}_c'(k) + k^2\mathcal{R}_c(k) = 0, \tag{14}$$

where $z \equiv a\sqrt{2\epsilon}$, k is the comoving wave number, and primes denote derivatives with respect to conformal time.

The Gaussianity of temperature fluctuations in the CMB implies that the statistical properties of the temperature field can be completely characterized by its two-point correlation function, i.e. its temperature spectrum [7, 9]. In the case of curvature perturbations its two-point correlation function is given by [7, 9]

$$\left\langle \hat{\mathcal{R}}_c(\vec{k}_1,t)\hat{\mathcal{R}}_c(\vec{k}_2,t)\right\rangle \equiv (2\pi)^3\frac{2\pi^2}{k^3}P_{\mathcal{R}_c}(k)\delta^{(3)}(\vec{k}_1+\vec{k}_2)\,, \tag{15}$$

where the power spectrum of primordial curvature perturbations is defined as

$$P_{\mathcal{R}_c}(k) \equiv \frac{2k^3}{(2\pi)^2}|\mathcal{R}_c(k)|^2. \tag{16}$$

We solve Eq. (14) numerically to obtain the spectrum of primordial curvature perturbations for the LF and BF. The different effects on the primordial spectrum are shown in figures 4 and 5. In figure 4 it can be seen that LF and BF produce similar oscillations of the spectrum around k_0 [15, 18, 19, 33–35, 51–53]. Nonetheless in figure 5 we can see that BF have steps at large and small scales, i.e. differences between the feature and featureless spectra at large and small

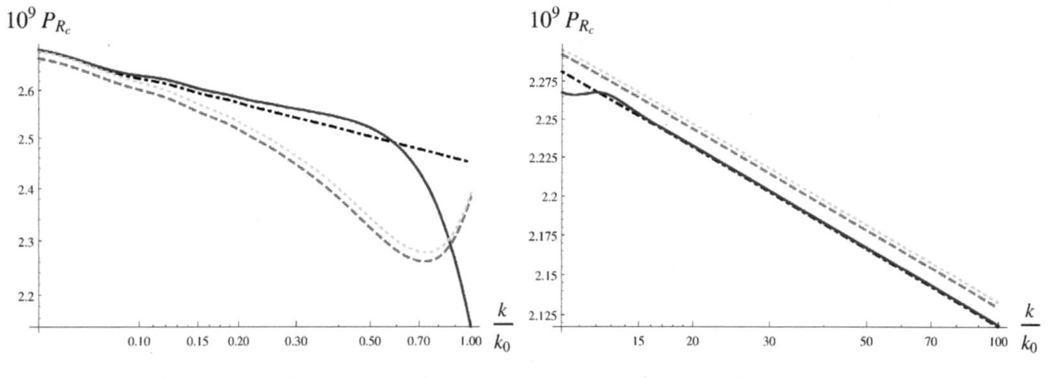

Figure 5. The numerically computed power spectrum of primordial curvature perturbations $P_{\mathcal{R}_c}$ is plotted for the potentials in Eqs. (5-7). In all cases we use $\gamma = 1.66\times10^{-11}, \sigma = 0.04$, and we set the scale of the feature at $k_0 = 1.5\times10^{-3}\text{Mpc}^{-1}$. For V_1, V_2, and V_3 we use $\lambda = -7\times10^{-12}$ (blue), $\lambda = -5.25\times10^{-12}$ (red-dashed), and $\lambda = -1.2\times10^{-3}$ (green-dotted), respectively. The dashed-dotted black lines correspond to the featureless spectrum. On the left (right) panel we make a zoom of the spectrum for large (small) scales.

scales, which are absent for the LF [35]. This indicates that the BF effects on the primordial perturbations are not only visible around the scale k_0 leaving the horizon around the feature time τ_0, but for any other scale leaving the horizon when ϕ has a value within the feature branch [15, 18, 19, 33–35, 51–53]. On the contrary LF only affect the perturbation modes which leave the horizon around τ_0, and consequently the spectrum does not show a step, but a local dumped oscillation and then it approaches the featureless spectrum for sufficiently smaller and larger scales [35]. This is crucial because it could allow to model local features of the temperature spectrum without affecting other scales [36]. Moreover we could distinguish between LF and BF by the predictions of their effects on the temperature spectrum at large and small scales, where they are quantitatively different.

For the parameters λ, σ and k_0 chosen, another important characteristic of the spectra obtained from V_2 and V_3 is that their behavior is very similar. In fact, from figure 4 we can see that the difference between the two spectra is quantitatively small. Only when we look at higher or smaller scales we can see a difference between them. If this difference is too small both primordial spectra could predict an indistinguishable CMB temperature spectrum which would make almost impossible to distinguish between the two different models.

It is also possible that this degeneracy could be easily broken at higher order terms, for instance by the three-point correlation function (or *bispectrum*) [55,56]. Implying the importance of studying higher order correlation functions to gain more insight about the physics of the very early Universe.

4. Conclusions

Future CMB and LSS surveys will offer an impressive opportunity for the discovery of primordial features over the next decades. They would give us profound insight into the physics of the primordial Universe such as probing the nature of inflationary models, distinguishing between inflation and alternative scenarios, and even discovering new particles.

In this paper we have studied the effects of LF and BF of the inflaton potential on the spectrum of primordial curvature perturbations. We have seen that LF affect the spectrum in a narrow range of scales, as opposed to BF which produce a step with respect to the featureless spectrum at large and small scales. It would be interesting to study if the differences between the LF and BF spectra at large and small scales can actually be quantified in a CMB spectrum analysis.

Another important result is the similarity between the primordial spectra predicted by the two BF potentials considered. This could lead to a degeneracy of the CMB temperature spectrum which could make even harder the discrimination between the models from a CMB analysis. It is possible though that the degeneracy can be broken when higher order perturbation terms are considered, such as the bispectrum. In this sense non-Gaussianity may play an important role in discerning between different inflationary models which predict similar spectra. It is left for a future work the study of such a degeneracy breaking for the models considered in this paper.

As a final remark, it would be interesting to study phenomenologically a singled potential containing both LF and BF and study its predictions on the primordial and CMB spectra.

Acknowledgments
This work was supported by the Colombian Department of Science, Technology, and Innovation COLCIENCIAS research Grant No. 617-2013. A.G.C. acknowledges the partial support from the International Center for Relativistic Astrophysics Network ICRANet during his stay in Italy and valuable comments from Clément Stahl.

References

[1] Peiris H *et al.* (WMAP Collaboration) 2003 *Astrophys.J.Suppl.* **148** 213–231 (*Preprint* astro-ph/0302225)
[2] Larson D, Dunkley J, Hinshaw G, Komatsu E, Nolta M *et al.* 2011 *Astrophys.J.Suppl.* **192** 16 (*Preprint* 1001.4635)
[3] Hinshaw G *et al.* (WMAP) 2013 *Astrophys.J.Suppl.* **208** 19 (*Preprint* 1212.5226)
[4] Bennett C *et al.* (WMAP) 2013 *Astrophys.J.Suppl.* **208** 20 (*Preprint* 1212.5225)
[5] Ade P A R *et al.* (Planck) 2014 *Astron. Astrophys.* **571** A1 (*Preprint* 1303.5062)
[6] Ade P A R *et al.* (Planck) 2014 *Astron. Astrophys.* **571** A16 (*Preprint* 1303.5076)
[7] Ade P *et al.* (Planck Collaboration) 2013 (*Preprint* 1303.5082)
[8] Ade P A R *et al.* (Planck) 2016 *Astron. Astrophys.* **594** A13 (*Preprint* 1502.01589)
[9] Ade P A R *et al.* (Planck) 2016 *Astron. Astrophys.* **594** A20 (*Preprint* 1502.02114)
[10] Adelman-McCarthy J K *et al.* (SDSS Collaboration) 2008 *Astrophys.J.Suppl.* **175** 297–313 (*Preprint* 0707.3413)
[11] NASA Cmb experiments URL https://lambda.gsfc.nasa.gov/product/expt/
[12] Adam R *et al.* (Planck) 2015 (*Preprint* 1502.01582)
[13] Guth A H 1981 *Phys.Rev.* **D23** 347–356
[14] Linde A D 1982 *Phys.Lett.* **B108** 389–393
[15] Starobinsky A A 1998 *Grav.Cosmol.* **4** 88–99 (*Preprint* astro-ph/9811360)
[16] Martin J, Ringeval C and Vennin V 2013 (*Preprint* 1303.3787)
[17] Leach S M, Sasaki M, Wands D and Liddle A R 2001 *Phys. Rev.* **D64** 023512 (*Preprint* astro-ph/0101406)
[18] Starobinsky A A 1992 *JETP Lett.* **55** 489–494
[19] Adams J A, Cresswell B and Easther R 2001 *Phys. Rev.* **D64** 123514 (*Preprint* astro-ph/0102236)
[20] Chluba J, Hamann J and Patil S P 2015 *Int. J. Mod. Phys.* **D24** 1530023 (*Preprint* 1505.01834)
[21] Gariazzo S, Lopez-Honorez L and Mena O 2015 *Phys. Rev.* **D92** 063510 (*Preprint* 1506.05251)
[22] Mooij S, Palma G A, Panotopoulos G and Soto A 2015 *JCAP* **1510** 062 [Erratum: JCAP1602,no.02,E01(2016)] (*Preprint* 1507.08481)
[23] Appleby S, Gong J O, Hazra D K, Shafieloo A and Sypsas S 2016 *Phys. Lett.* **B760** 297–301 (*Preprint* 1512.08977)
[24] Hunt P and Sarkar S 2015 *JCAP* **1512** 052 (*Preprint* 1510.03338)
[25] Ashoorioon A and Krause A 2006 (*Preprint* hep-th/0607001)
[26] Ashoorioon A, Krause A and Turzynski K 2009 *JCAP* **0902** 014 (*Preprint* 0810.4660)
[27] Ballardini M, Finelli F, Fedeli C and Moscardini L 2016 *JCAP* **1610** 041 (*Preprint* 1606.03747)
[28] Hazra D K, Shafieloo A, Smoot G F and Starobinsky A A 2016 *JCAP* **1609** 009 (*Preprint* 1605.02106)
[29] Dorn S, Ramirez E, Kunze K E, Hofmann S and Ensslin T A 2014 *JCAP* **1406** 048 (*Preprint* 1403.5067)
[30] Gariazzo S, Giunti C and Laveder M 2015 *JCAP* **1504** 023 (*Preprint* 1412.7405)
[31] Nicholson G, Contaldi C R and Paykari P 2010 *JCAP* **1001** 016 (*Preprint* 0909.5092)
[32] Hunt P and Sarkar S 2014 *JCAP* **1401** 025 (*Preprint* 1308.2317)
[33] Arroja F, Romano A E and Sasaki M 2011 *Phys.Rev.* **D84** 123503 (*Preprint* 1106.5384)

[34] Cadavid A G and Romano A E 2015 *Eur. Phys. J.* **C75** 589 (*Preprint* 1404.2985)
[35] Cadavid A G, Romano A E and Gariazzo S 2016 *Eur. Phys. J.* **C76** 385 (*Preprint* 1508.05687)
[36] Gallego Cadavid A, Romano A E and Gariazzo S 2016 (*Preprint* 1612.03490)
[37] Gallego Cadavid A and Romano A E 2015 *Nucl. Part. Phys. Proc.* **267-269** 254–256
[38] Benetti M and Alcaniz J S 2016 *Phys. Rev.* **D94** 023526 (*Preprint* 1604.08156)
[39] Xu Y, Hamann J and Chen X 2016 *Phys. Rev.* **D94** 123518 (*Preprint* 1607.00817)
[40] Chen X, Dvorkin C, Huang Z, Namjoo M H and Verde L 2016 *JCAP* **1611** 014 (*Preprint* 1605.09365)
[41] Di Valentino E, Gariazzo S, Gerbino M, Giusarma E and Mena O 2016 *Phys. Rev.* **D93** 083523 (*Preprint* 1601.07557)
[42] Palma G A 2015 *JCAP* **1504** 035 (*Preprint* 1412.5615)
[43] Martin J, Sriramkumar L and Hazra D K 2014 *JCAP* **1409** 039 (*Preprint* 1404.6093)
[44] Motohashi H and Hu W 2015 (*Preprint* 1503.04810)
[45] Hamann J, Covi L, Melchiorri A and Slosar A 2007 *Phys.Rev.* **D76** 023503 (*Preprint* astro-ph/0701380)
[46] Hazra D K, Aich M, Jain R K, Sriramkumar L and Souradeep T 2010 *JCAP* **1010** 008 (*Preprint* 1005.2175)
[47] Hazra D K, Shafieloo A and Souradeep T 2014 *JCAP* **1411** 011 (*Preprint* 1406.4827)
[48] Joy M, Sahni V and Starobinsky A A 2008 *Phys.Rev.* **D77** 023514 (*Preprint* 0711.1585)
[49] Joy M, Shafieloo A, Sahni V and Starobinsky A A 2009 *JCAP* **0906** 028 (*Preprint* 0807.3334)
[50] Mortonson M J, Dvorkin C, Peiris H V and Hu W 2009 *Phys.Rev.* **D79** 103519 (*Preprint* 0903.4920)
[51] Adshead P, Dvorkin C, Hu W and Lim E A 2012 *Phys.Rev.* **D85** 023531 (*Preprint* 1110.3050)
[52] Chen X, Easther R and Lim E A 2007 *JCAP* **0706** 023 (*Preprint* astro-ph/0611645)
[53] Hazra D K, Shafieloo A, Smoot G F and Starobinsky A A 2014 *JCAP* **1408** 048 (*Preprint* 1405.2012)
[54] Maldacena J M 2003 *JHEP* **0305** 013 (*Preprint* astro-ph/0210603)
[55] Ade P A R *et al.* (Planck) 2014 *Astron. Astrophys.* **571** A24 (*Preprint* 1303.5084)
[56] Ade P A R *et al.* (Planck) 2016 *Astron. Astrophys.* **594** A17 (*Preprint* 1502.01592)

IOP Conf. Series: Journal of Physics: Conf. Series **831** (2017) 012004 doi:10.1088/1742-6596/831/1/012004

PI/UAN-2017-601FT

Scalar and vector Galileons

Yeinzon Rodríguez[1,2,3] **and Andrés A. Navarro**[2]

[1] Centro de Investigaciones en Ciencias Básicas y Aplicadas, Universidad Antonio Nariño, Cra 3 Este # 47A-15, Bogotá D.C. 110231, Colombia

[2] Escuela de Física, Universidad Industrial de Santander, Ciudad Universitaria, Bucaramanga 680002, Colombia

[3] Simons Associate at The Abdus Salam International Centre for Theoretical Physics, Strada Costiera 11, I-34151, Trieste, Italy

E-mail: yeinzon.rodriguez@uan.edu.co, navarandres@gmail.com

Abstract. An alternative for the construction of fundamental theories is the introduction of Galileons. These are fields whose action leads to non higher than second-order equations of motion. As this is a necessary but not sufficient condition to make the Hamiltonian bounded from below, as long as the action is not degenerate, the Galileon construction is a way to avoid pathologies both at the classical and quantum levels. Galileon actions are, therefore, of great interest in many branches of physics, specially in high energy physics and cosmology. This proceedings contribution presents the generalities of the construction of both scalar and vector Galileons following two different but complimentary routes.

1. Introduction

A top-down approach to fundamental physics involves three steps: 1. we have to define the matter content we are interested in, 2. we have to establish what symmetries the action is going to enjoy, 3. we have to fix things here and there (removing some terms in the action, establishing relations among different coupling constants, etc.) so that no pathologies are present. This approach is extremely successful with remarkable examples such as the construction of the Standard Model of particle physics, Supersymmetry, and Supergravity. A complimentary approach consists in formulating the following question: is there any choice at constructing the fundamental theory? We can rephrase this question by asking if the removal of any sort of pathologies may lead to a unique action once the matter content and the action symmetries are defined. When talking about pathologies, we can start with the Ostrogradski's instability [1], that one where the state of the system goes down to lower and lower energy levels because the Hamiltonian is not bounded from below. Such an instability would lead to disastrous consequences both at the classical and quantum levels [2, 3]. Ostrogradski's theorem states that, as long as the action is not degenerate, equations of motion higher than second order lead to a Hamiltonian unbounded from below (for a review, see Refs. [2, 3]). Thus, if we want a healthy fundamental theory, the action has to be constructed so that equations of motion are second-order (at most). This indeed supports the fact that most of the physics laws are described by second-order differential equations (think, for example, of the Newton laws of mechanics, the Maxwell laws of electromagnetism, and the Einstein equations of gravity). Of course, this is not a sufficient condition for a healthy theory, so that a Hamiltonian analysis has to be performed once the action has been built.

In the mid 70's, G. W. Hordenski wrote the most general action for a scalar field and classical gravity that leads to non higher than second-order equations of motion [4]. A couple of years later, Horndeski himself did the same work exchanging the scalar field for an Abelian vector field [5]. His results were largely ignored until the first years of the 2010 decade when they were rediscovered [6, 7] in the framework of what is nowadays called Galileons [8]. A Galileon is a scalar field π whose action in flat spacetime enjoys a "Galilean" symmetry $\pi \to \pi + b_\mu x^\mu + c$, where b_μ and c are a constant four-vector and a constant scalar respectively, and whose equation of motion is not higher than second order. It can be shown that this Galilean symmetry is the responsible for the existence of an equation of motion involving strictly second-order space-time derivatives [8]. A generalization of the Galileon field destroys the Galilean symmetry but allows for lower than second-order space-time derivatives in the equations of motion both in flat and curved spacetime [9, 6]. Such a generalized Galileon is what we will call in this proceedings contribution, from now on, simply as a Galileon. Since the rediscovery of this idea, its consequences in high energy physics and cosmology have been subject of intensive study.

Life, nonetheless, does not end with scalar fields. Vector fields are also relevant players in high energy physics and its consequences in cosmology have been under scrutiny in the latest years (see, for example, Refs. [10, 11]). As stated before, Horndeski had studied what we can call now an "Abelian vector Galileon": a vector field whose action is invariant under the $U(1)$ group of symmetries. In flat spacetime, the only possibility is the Einstein-Hilbert-Maxwell action [12], but things change in curved spacetime [5]. As a step towards the fundamental theory, we can relax the gauge-invariance requirement, generalizing this way the Proca action. It was L. Heisenberg and G. Tasinato, independently, who explored this possibility for the first time [13, 14]. By following the Galileon construction method described in Refs. [9, 6], these authors built the generalized Proca action by contracting two Levi-Civita tensors with first-order derivatives $\partial_\mu A_\nu$ of the vector field A_μ. They obtained, this way, a theory where only three degrees of freedom associated to A_μ propagate. Moreover, the longitudinal degree of freedom behaves as a scalar Galileon, which means that the theory has a safe decoupling limit. Such a procedure is powerful but has its limitation: because of the contraction with the two Levi-Civita tensors, it only produces parity-conserving terms, precluding the construction of allowed parity-violating interactions. By following a complimentary route, E. Allys, P. Peter and Y. Rodríguez built all the possible Lorentz-invariant terms involving A_μ and $\partial_\mu A_\nu$ and wrote linear combinations consistent with the propagation of three degrees of freedom for the vector field and a safe decoupling limit [15]. The limitation of this procedure is that it does not provide a clear limit in the construction in the sense of a threshold in the number of first-order derivatives of A_μ in each term of the action. In contrast, its advantage is that it does generate parity-violating terms. Some controversy around these results was initiated with the publication of Ref. [16] but a final agreement was reached in Ref. [17].

The purpose of this paper is to show, schematically, the two Galileon construction procedures described in the previous paragraph, for both a scalar and vector fields. For the latter, I will show its application to an Abelian vector field and to a set of $SU(2)$ gauge fields. A short description of the advances in the study of the cosmological implications of the $SU(2)$ vector Galileons will be given at the end.

2. Scalar Galileon

Let's assume a scalar field π in flat spacetime whose action is Lorentz invariant. The action may contain the field itself and the first-order and second-order space-time derivatives $\partial_\mu \pi$ and $\partial_\mu \partial_\nu \pi$. Any higher than second-order space-time derivative would lead, immediately, to higher than second-order equations of motion. Thus, the action being $S = \int \mathcal{L}(\pi, \partial_\mu \pi, \partial_\mu \partial_\nu \pi)\, d^4x$, the

Euler-Lagrange equations become

$$\frac{\partial \mathcal{L}}{\partial \pi} - \partial_\mu \frac{\partial \mathcal{L}}{\partial(\partial_\mu \pi)} + \partial_\mu \partial_\nu \frac{\partial \mathcal{L}}{\partial(\partial_\mu \partial_\nu \pi)} = 0\,, \tag{1}$$

which shows that special care has to be taken with those terms in $\mathcal{L}$ involving second-order space-time derivatives.

The strategy to follow, then, to build the most general action leading to non higher than second-order equations of motion consists in

(i) identifying all the possible Lorentz-invariant terms built from contractions of first-order and second-order space-time derivatives with metric tensors,

(ii) grouping together all these Lorentz-invariant terms in general linear combinations,

(iii) establishing relations among the coefficients in the linear combinations so that the higher than second-order contributions produced by the third term in Eq. (1) vanish.

As can be easily checked, the most general action is just given by these linear combinations multiplied by arbitrary functions of π.

So, what is the structure of these linear combinations? A careful study of the Euler-Lagrange equations, following the strategy described above, reveals that the adequate combination of Lorentz-invariant terms in the Lagrangian density comes in four different pieces $\mathcal{L}^{\rm Gal}_{N,\pi}$ starting from $N = 2$ [9, 6]:

$$\mathcal{L}^{\rm Gal}_{N,\pi} \equiv f_N(\pi, X)\, \frac{1}{(4-n)!} \epsilon^{\mu_1 \ldots \mu_n \sigma_1 \ldots \sigma_{4-n}} \epsilon^{\nu_1 \ldots \nu_n}{}_{\sigma_1 \ldots \sigma_{4-n}} \, (\partial_{\mu_1} \partial_{\nu_1} \pi) \ldots (\partial_{\mu_n} \partial_{\nu_n} \pi)\,, \tag{2}$$

where $N \equiv n + 2$, the ϵ tensors are Levi-Civita, and the $f_N(\pi, X)$ are arbitrary functions of π and $X \equiv \partial_\lambda \pi \partial^\lambda \pi$. Being explicit, the action is written this way:

$$S = \int \sum_{N=2}^{5} \mathcal{L}^{\rm Gal}_{N,\pi}\, d^4x\,, \tag{3}$$

where

$$\mathcal{L}^{\rm Gal}_{2,\pi} \equiv f_2(\pi, X)\,, \tag{4}$$
$$\mathcal{L}^{\rm Gal}_{3,\pi} \equiv f_3(\pi, X)\, \Box \pi\,, \tag{5}$$
$$\mathcal{L}^{\rm Gal}_{4,\pi} \equiv f_4(\pi, X)\, [(\Box \pi)^2 - (\partial_\mu \partial_\nu \pi)(\partial^\mu \partial^\nu \pi)]\,, \tag{6}$$
$$\mathcal{L}^{\rm Gal}_{5,\pi} \equiv f_5(\pi, X)\, [(\Box \pi)^3 - 3(\Box \pi)(\partial_\mu \partial_\nu \pi)(\partial^\mu \partial^\nu \pi) + 2(\partial_\mu \partial^\nu \pi\; \partial_\nu \partial^\rho \pi\; \partial_\rho \partial^\mu \pi)]\,. \tag{7}$$

For applications in curved spacetime, the partial derivatives must be replaced by covariant derivatives which, unfortunately, introduce contributions to the equations of motion involving higher than second-order space-time derivatives of both the scalar field and the metric [18, 19]. This issue is easily solved by introducing specific counterterms in $\mathcal{L}^{\rm Gal}_{4,\pi}$ and $\mathcal{L}^{\rm Gal}_{5,\pi}$ [6]. The action is now written this way:

$$S = \int \left[\sum_{N=2}^{5} \mathcal{L}^{\rm Gal}_{N,\pi} + \mathcal{L}^{\rm Gal}_{{\rm Curv},\pi} \right] \sqrt{-g}\, d^4x\,, \tag{8}$$

IOP Conf. Series: Journal of Physics: Conf. Series **831** (2017) 012004 doi:10.1088/1742-6596/831/1/012004

where

$$\mathcal{L}^{\rm Gal}_{{\rm Curv},\pi} \equiv G^{\rm Curv}(\pi)\, G_{\mu\nu}(\nabla^\mu\pi\nabla^\nu\pi)\,, \tag{9}$$

$$\mathcal{L}^{\rm Gal}_{2,\pi} \equiv G_2(\pi, X)\,, \tag{10}$$

$$\mathcal{L}^{\rm Gal}_{3,\pi} \equiv G_3(\pi, X)\,\Box\pi\,, \tag{11}$$

$$\mathcal{L}^{\rm Gal}_{4,\pi} \equiv G_4(\pi, X)R + G_{4,X}\,[(\Box\pi)^2 - (\nabla_\mu\nabla_\nu\pi)(\nabla^\mu\nabla^\nu\pi)]\,, \tag{12}$$

$$\begin{aligned}\mathcal{L}^{\rm Gal}_{5,\pi} \equiv\ & G_5(\pi, X)G_{\mu\nu}(\nabla^\mu\nabla^\nu\pi) \\ & -\frac{1}{6}G_{5,X}\,[(\Box\pi)^3 - 3(\Box\pi)(\nabla_\mu\nabla_\nu\pi)(\nabla^\mu\nabla^\nu\pi) + 2(\nabla_\mu\nabla^\nu\pi\ \nabla_\nu\nabla^\rho\pi\ \nabla_\rho\nabla^\mu\pi)]\,,\end{aligned} \tag{13}$$

where g is the determinant of the metric, R is the Ricci scalar, $G_{\mu\nu}$ is the Einstein tensor, ∇ stands for a covariant derivative, and $G_{N,X} \equiv \partial G_N/\partial X$. As we can see, a new term is introduced, $\mathcal{L}^{\rm Gal}_{{\rm Curv},\pi}$, which vanish in flat spacetime; since $G_{\mu\nu}$ is divergenceless, this new term, likewise the counterterms in $\mathcal{L}^{\rm Gal}_{4,\pi}$ and $\mathcal{L}^{\rm Gal}_{5,\pi}$, does not introduce new propagating degrees of freedom in the tensor sector [5, 20].

It is clear that Eq. (2) is just a consequence of examining the Euler-Lagrange equations in the search for non higher than second-order equations of motion. Following the strategy described above or employing directly Eq. (2) are, therefore, equivalent procedures to build the action for a scalar Galileon. Such an equivalence is not necessarily present, however, when constructing vector Galileons, as the next section shows.

3. Vector Galileons

3.1. The generalized Abelian Proca action

Let's start in flat spacetime. Because of the Helmholtz decomposition, any space-time vector A_μ can be split in two pieces:

$$A_\mu = \mathcal{A}_\mu + \partial_\mu\pi\,, \tag{14}$$

where $\mathcal{A}_\mu$ is divergenceless and π is a scalar field. π is, therefore, the longitudinal degree of freedom associated to A_μ. Thus, should A_μ be identified as a vector Galileon, π must be a scalar Galileon. The immediate consequence of this fact is that the action of the vector field may include the field itself and its first-order space-time derivative $\partial_\mu A_\nu$ only. This is welcome since the equation of motion for A_μ will not be higher than second order no matter the specific form of the action.

The first route to build the Galilean action follows similar lines to those described for the scalar Galileon case:

(i) identify all the possible Lorentz-invariant terms built from contractions of vector fields and first-order space-time derivatives with the primitive invariants of the Lorentz group ($SO(3,1)$): metric tensors and, at most, one Levi-Civita tensor,

(ii) group together all these Lorentz-invariant terms in general linear combinations,

(iii) establish relations among the coefficients in the linear combinations so that no more than three degrees of freedom propagate,

(iv) make the replacement $A_\mu \to \partial_\mu\pi$ and remove all those terms whose resultant action does not correspond to that of a scalar Galileon.

The restriction to the number of Levi-Civita tensors in the first step comes from the fact that the product of two such tensors can always be expressed as a linear combination of products of metric tensors. Notice that, because of the antisymmetric properties of the Levi-Civita tensor, the analogous step for a scalar Galileon only requires contractions with metric tensors. Regarding the third step, the propagation of just three degrees of freedom, they being the spatial components of the vector field, is established by the Hessian condition $\mathcal{H}^{0\nu} = 0$ [13] where

$$\mathcal{H}^{\mu\nu} \equiv \frac{\partial^2 \mathcal{L}}{\partial(\partial_0 A_\mu)\partial(\partial_0 A_\nu)} . \tag{15}$$

Thus, the action for the vector Galileon, which generalizes the Abelian Proca action, turns out to be

$$S = \int \left[-\frac{1}{4} F_{\mu\nu} F^{\mu\nu} + \frac{1}{2} m^2 A^2 + \sum_{N=2}^{6} \mathcal{L}_{N,A}^{\rm Gal} \right] d^4x , \tag{16}$$

where [16, 17]

$$\begin{aligned}
\mathcal{L}_{2,A}^{\rm Gal} &\equiv f_2(A_\mu, F_{\mu\nu}, \tilde{F}_{\mu\nu}) = f_2 \left[A^2, F^2, F \cdot \tilde{F}, (A \cdot \tilde{F})^2 \right] , &(17)\\
\mathcal{L}_{3,A}^{\rm Gal} &\equiv f_3(A^2) \; S^\mu{}_\mu , &(18)\\
\mathcal{L}_{4,A}^{\rm Gal} &\equiv f_4(A^2) \; [(S^\mu{}_\mu)^2 - S_\rho{}^\sigma S_\sigma{}^\rho] , &(19)\\
\mathcal{L}_{5,A}^{\rm Gal} &\equiv f_5(A^2) \; [(S^\mu{}_\mu)^3 - 3(S^\mu{}_\mu) S_\rho{}^\sigma S_\sigma{}^\rho + 2 S_\rho{}^\sigma S_\sigma{}^\gamma S_\gamma{}^\rho] + g_5(A^2) \; \tilde{F}^{\alpha\mu} \tilde{F}^\beta{}_\mu S_{\alpha\beta} , &(20)\\
\mathcal{L}_{6,A}^{\rm Gal} &\equiv g_6(A^2) \; \tilde{F}^{\alpha\beta} \tilde{F}^{\mu\nu} S_{\alpha\mu} S_{\beta\nu} . &(21)
\end{aligned}$$

In contrast to the local gauge field theory, a new element appears in the action accompanying A_μ, the field strength tensor $F_{\mu\nu} \equiv \partial_\mu A_\nu - \partial_\nu A_\mu$, and its Hodge dual $\tilde{F}_{\mu\nu} \equiv \frac{1}{2}\epsilon_{\mu\nu\rho\sigma} F^{\rho\sigma}$. This new element is the symmetric version of the field strength tensor: $S_{\mu\nu} \equiv \partial_\mu A_\nu + \partial_\nu A_\mu$. It is important to notice that the Lagrangian pieces whose coefficient is a function f_N reduce to the respective scalar Galileon Lagrangians of Eqs. (4) - (7) when A_μ is replaced by $\partial_\mu \pi$ (except for the possible π dependence of f_N). Meanwhile, those Lagrangian pieces premultiplied by a function g_N vanish when we make the same replacement; this is why the generalized Proca action has an additional Lagrangian, $\mathcal{L}_{6,A}^{\rm Gal}$, compared to its scalar counterpart, which was not discovered in the pioneering papers of L. Heisenberg and G. Tasinato [13, 14].

This way of procedure can be extended for an arbitrary number of first-order derivatives without a complete proof why the sequence should stop. This indeed led to E. Allys, P. Peter, and Y. Rodríguez, in Ref. [15], to construct all the Lagrangian pieces up to $\mathcal{L}_{7,A}^{\rm Gal}$ and to conjecture that an infinite tower of terms would be generated; however, as shown in Ref. [16], the $\mathcal{L}_{7,A}^{\rm Gal}$ Lagrangian presented in Ref. [15] identically vanishes. This is an unfortunate limitation of the procedure already described which is not present in the second route we will present shortly. The advantage of the method already described is that it generates parity-violating terms: since some of the Lorentz-invariant terms in the first step are built from contractions with just one Levi-Civita tensor, parity-violating terms are inevitable. Most of these terms end up encoded in $\mathcal{L}_{2,A}^{\rm Gal}$ while the others vanish identically (at least, up to $\mathcal{L}_{7,A}^{\rm Gal}$) [17]. There exists a special parity-violating term uncovered in Ref. [17] which is impossible to build following the second route. This term is specially different to the other terms in the Lagrangian because not all the A_μ pairs come as A^2:

$$\mathcal{L}_{4,A}^{\rm Gal,bis} \equiv g_4(A^2) \; A^\mu \tilde{F}_{\mu\nu} S^{\nu\lambda} A_\lambda . \tag{22}$$

This term, however, is redundant (it already belongs to $\mathcal{L}_{2,A}^{\rm Gal}$) since $\tilde{F}_{\mu\nu}$ is divergenceless (see the Appendix).

The curved spacetime version of the generalized Proca action is obtained, as in the scalar Galileon case, by replacing the standard space-time derivatives by covariant derivatives and adding the required counterterms. The latter are chosen in order to avoid higher than second-order equations of motion and the wrong number of propagating degrees of freedom at every level of the decoupling limit. Refs. [13, 15, 16] arrive to the following result:

$$S = \int \left[-\frac{1}{4}F_{\mu\nu}F^{\mu\nu} + \frac{1}{2}m^2A^2 + \sum_{N=2}^{6} \mathcal{L}^{\rm Gal}_{N,A} + \mathcal{L}^{\rm Gal}_{{\rm Curv},A} \right] \sqrt{-g}\, d^4x \,, \tag{23}$$

where

$$\mathcal{L}^{\rm Gal}_{{\rm Curv},A} \equiv f^{\rm Curv} G_{\mu\nu}A^\mu A^\nu \,, \tag{24}$$

$$\mathcal{L}^{\rm Gal}_{2,A} \equiv f_2(A_\mu, F_{\mu\nu}, \tilde{F}_{\mu\nu}) = f_2\left[A^2, F^2, F\cdot\tilde{F}, (A\cdot\tilde{F})^2\right] \,, \tag{25}$$

$$\mathcal{L}^{\rm Gal}_{3,A} \equiv f_3(A^2)\, S^\mu{}_\mu \,, \tag{26}$$

$$\mathcal{L}^{\rm Gal}_{4,A} \equiv f_4(A^2)R + \frac{1}{2}f_{4,A^2}\,[(S^\mu{}_\mu)^2 - S_\rho{}^\sigma S_\sigma{}^\rho] \,, \tag{27}$$

$$\mathcal{L}^{\rm Gal}_{5,A} \equiv f_5(A^2)G_{\mu\nu}S^{\mu\nu} + \frac{3}{4}f_{5,A^2}\,[(S^\mu{}_\mu)^3 - 3(S^\mu{}_\mu)S_\rho{}^\sigma S_\sigma{}^\rho + 2S_\rho{}^\sigma S_\sigma{}^\gamma S_\gamma{}^\rho] + g_5(A^2)\,\tilde{F}^{\alpha\mu}\tilde{F}^\beta{}_\mu S_{\alpha\beta} \,, \tag{28}$$

$$\mathcal{L}^{\rm Gal}_{6,A} \equiv g_6(A^2)L_{\mu\nu\rho\sigma}F^{\mu\nu}F^{\rho\sigma} + g_{6,A^2}\,\tilde{F}^{\alpha\beta}\tilde{F}^{\mu\nu}S_{\alpha\mu}S_{\beta\nu} \,, \tag{29}$$

where $f^{\rm Curv}$ is just a constant scalar and $L_{\mu\nu\rho\sigma}$ is the double dual Riemann tensor:

$$L^{\mu\nu\alpha\beta} \equiv \frac{1}{4}\epsilon^{\mu\nu\rho\sigma}\epsilon^{\alpha\beta\gamma\delta}R_{\rho\sigma\gamma\delta} \,, \tag{30}$$

$R_{\rho\sigma\gamma\delta}$ being the Riemann tensor.

The second route to build the generalized Proca action consists in extending Eq. (2), valid for a scalar Galileon, to the vector field A_μ. This means contracting products of first-order space-time derivatives of A_μ with two Levi-Civita tensors:

$$\mathcal{L}^{\rm Gal}_{N,A} \propto \epsilon^{-}{}_{-}\epsilon^{-}{}_{-}\ \partial_{\cdot}A_{\cdot}\partial_{\cdot}A_{\cdot}\cdots \,. \tag{31}$$

What is interesting about this construction is that, as shown in Ref. [17], the Hessian condition is satisfied automatically. In addition, when making the replacement $A_\mu \to \partial_\mu\pi$, Eq. (31) reduces to Eq. (2), except for $N = 6$, so that the theory has a safe decoupling limit (for $N = 6$, we have to discard that term that does not vanish when taking the scalar limit, see Ref. [16]). Eq. (31) reproduces Eqs. (17) - (21) except for the fact that Eq. (17) has to be understood, in this approach, as containing only parity-conserving terms. This is the limitation of this procedure: since the Lagrangian pieces are built from contractions with two Levi-Civita tensors, it is impossible to build parity-violating terms. There is however an even bigger limitation: there does not exist a proof that Eq. (31) generates all the possible Lagrangian pieces of the generalized Proca action in contrast with Eq. (2) for which there does exist a formal proof (see Ref. [6]). Since all the contractions of space-time derivatives of A_μ with the primitive invariants of the Lorentz groups can be written as contractions of the former with an unrestricted number of Levi-Civita tensors, there is no reason why the number of Levi-Civita tensors should be only two. Thus, although the second route gives a finite number of Lagrangian pieces, since the number of space-time indices in the Levi-Civita tensors contracted with space-time derivatives is saturated in $\mathcal{L}^{\rm Gal}_{6,A}$, it might happen that the actual number of Lagrangian pieces contributing to the generalized Proca action is larger.

IOP Conf. Series: Journal of Physics: Conf. Series **831** (2017) 012004 doi:10.1088/1742-6596/831/1/012004

3.2. The generalized SU(2) Proca action

Let's now consider a set of three vector fields A^a_μ, with a running from 1 to 3, whose action is invariant under the $SU(2)$ global symmetry group. Grouping these vector fields into a single matrix $\mathcal{A}_\mu = A^a_\mu T_a$, where the T_a are the matrix generators of the $SU(2)$ transformations, it is possible to show that $\mathcal{A}_\mu$ transforms in the respective adjoint representation (see, for example, Ref. [21]):

$$\mathcal{A}'_\mu = e^{ig\vec{\epsilon}\cdot\vec{T}}\mathcal{A}_\mu e^{-ig\vec{\epsilon}\cdot\vec{T}} \,. \tag{32}$$

In the previous expression, g is the coupling constant, $\vec{\epsilon}$ is a three-dimensional vector that parametrizes the amount of the transformation, and $\vec{T}$ is the "vector" built with the matrix generators. Such generators satisfy the Lie algebra

$$[T_a, T_b] = \epsilon_{ab}{}^{c}\, T_c \,, \tag{33}$$

where ϵ_{abc} denotes the structure constants of the group which, for $SU(2)$, correspond to the Levi-Civita symbol.

With these preliminaries in mind, the first route to build the generalized $SU(2)$ Proca action consists in the following steps:

(i) identify all the possible Lorentz-invariant terms built from contractions of vector fields A_μ (without $SU(2)$ group indices) and first-order space-time derivatives $\partial_\mu A_\nu$ with the primitive invariants of the Lorentz group ($SO(3,1)$): metric tensors and, at most, one Levi-Civita tensor,

(ii) add $SU(2)$ group indices to the terms identified in the previous step and contract with the primitive invariants of the $SU(2)$ group: group induced metrics g_{ab} and structure constants ϵ_{abc} (at most, one Levi-Civita symbol).

(iii) group together all these Lorentz-$SU(2)$ invariant terms in general linear combinations,

(iv) establish relations among the coefficients in the linear combinations so that no more than three degrees of freedom propagate,

(v) make the replacement $A^a_\mu \to \partial_\mu \pi^a$ and remove all those terms whose resultant action does not correspond to that of multi-scalar Galileons in the three-dimensional representation of $SU(2)$ (see Ref. [22] and Appendix A of Ref. [23]; see also Ref. [24]).

Because of symmetry reasons, some of the terms found in the first step vanish identically; however, they must be kept since the addition of group indices in the second step can make them non vanishing. The restriction to the number of Levi-Civita symbols in the second step comes from the fact that the product of two such symbols can always be expressed as a linear combination of products of group induced metrics [25].

This procedure was followed in Ref. [23] with the imposed restriction of, at most, six space-time indices in the Lorentz-$SU(2)$ invariant terms. Such a restriction is just technical in its nature: it was imposed because the number of Lorentz-$SU(2)$ invariant terms scales strongly when allowing for more indices, as Table 1 shows.

Thus, the flat spacetime generalized $SU(2)$ Proca action is given by

$$S = \int \left[-\frac{1}{4} F^a_{\mu\nu} F_a^{\mu\nu} + \frac{1}{2} m^2 A^\mu_a A^a_\mu + \sum_{N=2}^{4} \mathcal{L}^{\rm Gal}_{N,A} \right] d^4x \,, \tag{34}$$

where [23]

$$\mathcal{L}^{\rm Gal}_{2,A} \equiv f_2(A^a_\mu, F^a_{\mu\nu}, \tilde{F}^a_{\mu\nu}) \,, \tag{35}$$

$$\mathcal{L}^{\rm Gal}_{3,A} \equiv 0 \,, \tag{36}$$

Table 1. Number of Lorentz-$SU(2)$ invariant terms built from contractions of some number of vector fields (first line) and some number of first-order derivatives (first column) with the primitive invariants of $SO(3,1)$ and $SU(2)$.

$\#\partial^\mu A^{\nu a}$ \ $\#A^{\rho b}$	0	2	4
1	0	3	36
2	4	42	510
3	9	312	

$$\begin{aligned}\mathcal{L}_{4,A}^{\text{Gal}} \equiv\ & f_4^1 \Big\{(A_b \cdot A^b)\,[(\partial \cdot A_a)\,(\partial \cdot A^a) - (\partial_\mu A_a^\nu)(\partial^\mu A_\nu^a)] \\ & +2(A_a \cdot A_b)[(\partial \cdot A^a)\,(\partial \cdot A^b) - (\partial_\mu A^{\nu a})(\partial^\mu A_\nu^b)]\Big\} \\ & +f_4^2 \Big\{(A_a \cdot A_b)[(\partial \cdot A^a)\,(\partial \cdot A^b) - (\partial_\mu A^{\nu a})(\partial^\mu A_\nu^b)] \\ & +(A^{\mu a} A^{\nu b})\,[(\partial_\mu A_a^\alpha)\,(\partial_\nu A_{\alpha b}) - (\partial_\nu A_a^\alpha)\,(\partial_\mu A_{\alpha b})]\Big\} \\ & +f_4^3\, \tilde{G}_{\mu\sigma}^b A_a^\mu A_{\alpha b} S^{\alpha\sigma a}\,, \end{aligned} \tag{37}$$

where the f_4 are arbitrary constant scalars, $F_{\mu\nu}^a$ is the non-Abelian field strength tensor:

$$F_{\mu\nu}^a \equiv \partial_\mu A_\nu^a - \partial_\nu A_\nu^a + g\epsilon^a{}_{bc} A_\mu^b A_\nu^c\,, \tag{38}$$

$G_{\mu\nu}^a$ is the Abelian version of $F_{\mu\nu}^a$:

$$G_{\mu\nu}^a \equiv \partial_\mu A_\nu^a - \partial_\nu A_\nu^a\,, \tag{39}$$

and $S_{\mu\nu}^a$ is the symmetrized version of $G_{\mu\nu}^a$:

$$S_{\mu\nu}^a \equiv \partial_\mu A_\nu^a + \partial_\nu A_\nu^a\,. \tag{40}$$

Although $\mathcal{L}_{5,A}^{\text{Gal}}$ and $\mathcal{L}_{6,A}^{\text{Gal}}$ were not built, a comparison with the Abelian case allowed the authors of Ref. [23] to conjecture that

$$\mathcal{L}_{5,A}^{\text{Gal}} \equiv f_5\ \epsilon_{abc} \left(A^a \cdot A^d\right) \tilde{G}_d^{\alpha\mu} \tilde{G}^{\beta b}{}_\mu S_{\alpha\beta}^c\,, \tag{41}$$

and

$$\mathcal{L}_{6,A}^{\text{Gal}} \equiv f_6^1\ \tilde{G}_a^{\alpha\beta} \tilde{G}^{\mu\nu a} S_{\alpha\mu}^b S_{\beta\nu b} + f_6^2\ \tilde{G}_a^{\alpha\beta} \tilde{G}_b^{\mu\nu} S_{\alpha\mu}^a S_{\beta\nu}^b\,, \tag{42}$$

where f_5 and the f_6 are arbitrary constant scalars. It is very interesting to notice that the non-Abelian version of $\mathcal{L}_{4,A}^{\text{Gal,bis}}$, the last term in Eq. (37), is not a redundant term because the group indices protect it; this is, then, the first parity-violating term that is not included in f_2.

The curved spacetime version of this action is a bit more elaborate in its counterterms:

$$S = \int \left[-\frac{1}{4} F_{\mu\nu}^a F_a^{\mu\nu} + \frac{1}{2} m^2 A_a^\mu A_\mu^a + \sum_{N=2}^{4} \mathcal{L}_{N,A}^{\text{Gal}} + \sum_{m=1}^{5} \mathcal{L}_{\text{Curv},m,A}^{\text{Gal}}\right] \sqrt{-g}\ d^4x\,, \tag{43}$$

where [23]

$$\begin{aligned}
\mathcal{L}^{\text{Gal}}_{\text{Curv},1,A} &\equiv f_1^{\text{Curv}} G_{\mu\nu} A^{\mu a} A^{\nu}_a \,, & (44)\\
\mathcal{L}^{\text{Gal}}_{\text{Curv},2,A} &\equiv f_2^{\text{Curv}} L_{\mu\nu\rho\sigma} F^{\mu\nu a} F^{\rho\sigma}_a \,, & (45)\\
\mathcal{L}^{\text{Gal}}_{\text{Curv},3,A} &\equiv f_3^{\text{Curv}} \epsilon_{abc} L_{\mu\nu\rho\sigma} F^{\mu\nu a} A^{\rho b} A^{\sigma c} \,, & (46)\\
\mathcal{L}^{\text{Gal}}_{\text{Curv},4,A} &\equiv f_4^{\text{Curv}} L_{\mu\nu\rho\sigma} A^{\mu a} A^{\nu}_a A^{\rho b} A^{\sigma}_b \,, & (47)\\
\mathcal{L}^{\text{Gal}}_{2,A} &\equiv f_2(A^a_\mu, F^a_{\mu\nu}, \tilde{F}^a_{\mu\nu}) \,, & (48)\\
\mathcal{L}^{\text{Gal}}_{3,A} &\equiv 0 \,, & (49)\\
\mathcal{L}^{\text{Gal}}_{4,A} &\equiv f_4^1 \Big\{ (A_b \cdot A^b) \Big[(\nabla \cdot A_a)(\nabla \cdot A^a) - (\nabla_\mu A^\nu_a)(\nabla^\mu A^a_\nu) + \frac{1}{4}(A_a \cdot A^a) R \Big] \\
&+ 2(A_a \cdot A_b) \Big[(\nabla \cdot A^a)(\nabla \cdot A^b) - (\nabla_\mu A^{\nu a})(\nabla^\mu A^b_\nu) + \frac{1}{2}(A^a \cdot A^b) R \Big] \Big\} \\
&+ f_4^2 \Big\{ (A_a \cdot A_b) \Big[(\nabla \cdot A^a)(\nabla \cdot A^b) - (\nabla_\mu A^{\nu a})(\nabla^\mu A^b_\nu) + \frac{1}{4}(A^a \cdot A^b) R \Big] \\
&+ (A^{\mu a} A^{\nu b}) \Big[(\nabla_\mu A^\alpha_a)(\nabla_\nu A_{\alpha b}) - (\nabla_\nu A^\alpha_a)(\nabla_\mu A_{\alpha b}) - \frac{1}{2} A^\rho_a A^\sigma_b R_{\mu\nu\rho\sigma} \Big] \Big\} \\
&+ f_4^3 \, \tilde{G}^b_{\mu\sigma} A^\mu_a A_{\alpha b} S^{\alpha\sigma a} \,, & (50)
\end{aligned}$$

where the f^{Curv} are just constant scalars.

The second route to build the generalized $SU(2)$ Proca action is employed in Ref. [26]. Two Levi-Civita tensors are contracted with products of vector fields and their first-order derivatives:

$$\mathcal{L} \propto \epsilon^{\mu\nu\alpha\beta} \epsilon^{\rho\sigma\gamma\delta} \partial_\mu A^a_\rho \cdots A^b_\nu A^c_\sigma \cdots \,, \tag{51}$$

the free Lorentz indices being appropriately contracted with space-time metrics and the group indices being contracted with the primitive invariants of the $SU(2)$ group. Similarly to the Abelian case, this construction automatically satisfies the Hessian condition $\mathcal{H}^{0\nu ab} = 0$ where [23]

$$\mathcal{H}^{\mu\nu ab} \equiv \frac{\partial^2 \mathcal{L}}{\partial(\partial_0 A_{\mu a}) \partial(\partial_0 A_{\nu b})} \,, \tag{52}$$

so that just three degrees of freedom are able to propagate. In adition, the scalar limit has to be taken and checked in order to be sure that the theory has a safe decoupling limit. Unfortunately, Ref. [26] lacks for this last step. The same pros and contras of this route discussed in previous pages apply here as well. We stress that there does not exist a formal proof that applying Eq. (51) is equivalent to the first route described above; indeed, such a proof cannot exist due to the existence of the non redundant parity-violating term in $\mathcal{L}^{\text{Gal}}_{4,A}$ (the last one in Eq. (50)).

3.2.1. Cosmological implications We will now talk about some advances already made in the exploration of the cosmological implications of the action in Eq. (43) [27]. The idea is to have a "cosmic triad" configuration, i.e., the three vector fields orthogonal to each other and of the same norm, with the temporal components vanishing, in a Friedmann-Lemaitre-Robertson-Walker background. Such a configuration avoids any kind of anisotropy, both at the background and the perturbative level [28], which are severely restricted by the observations [29, 30, 31, 32]. It has also been studied previously with success in the Gauge-flation [33, 34, 35] and Chromo-natural inflation [36] models. We have discovered that all the Lagrangian pieces in Eq. (43) admit this configuration, except for the parity-violating term in the last line of Eq. (50) and,

IOP Conf. Series: Journal of Physics: Conf. Series **831** (2017) 012004 doi:10.1088/1742-6596/831/1/012004

in principle, some terms belonging to $\mathcal{L}_{2,A}^{\rm Gal}$. For its own existence, the parity-violating term requires not only at least one non vanishing temporal component but also a non orthogonal configuration. The Lagrangian in Eq. (41) shares the same requirements. In contrast, the Lagrangian in Eq. (42) admits the orthogonality but requires the existence of the temporal components; this implies momentum flow in the energy-momentum tensor which destroys the required isotropy. The Lagrangian in Eq. (45) had already been studied in Ref. [37], finding that a short period of inflation is generated; however, as shown in Ref. [38], this Lagrangian is plagued by ghosts and Laplacian instabilities. The other $\mathcal{L}_{\rm Curv,A}^{\rm Gal}$, except for $\mathcal{L}_{\rm Curv,1,A}^{\rm Gal}$ lead to very similar results at the background level and, probably, will have the same fate as $\mathcal{L}_{\rm Curv,2,A}^{\rm Gal}$ at the perturbative level. Regarding $\mathcal{L}_{\rm Curv,1,A}^{\rm Gal}$, it does not lead to any interesting cosmology at the background level. We are still trying to uncover the cosmological implications of the Lagrangian in Eq. (50) (except for the parity-violating term).

4. Conclusions

Two different but complimentary routes were exposed that lead to the construction of scalar and vector Galileon actions. Both routes are equivalent when building the Galileon action for a scalar field; however, such equivalence does not necessarily exist when building vector Galileons. Although the first route produces both parity-conserving and parity-violating terms, it does not give any clear end to the sequence of Lagrangian pieces that can be built. Indeed, it is possible, although unlikely, that such a sequence goes to infinity. In contrast, the second route does not produce parity-violating terms but it does provide a clear limit for the sequence of Lagrangians. Unfortunately, the second route is based on an unproved assumption. This does not preclude, anyway, the enormous help that the second route gives us to build and understand the structure of the Galileon actions. As a final remark, it is worth mentioning that the construction of the Galileon action involving simultaneously a scalar and a vector field is currently underway [39].

Acknowledgments

Y.R. wishes to thank E. Allys, P. Peter, and J. P. Beltrán Almeida for the nice collaboration they were involved that allowed him to immerse in the beautiful world of the Galileon/Horndeski theories. He also wishes to thank Ryo Namba and Lavinia Heisenberg for interesting and useful recent correspondence. We acknowledge Alejandro Guarnizo and Luis Gabriel Gómez for their help with the analysis of some terms in the generalized $SU(2)$ Proca action. This work was supported by COLCIENCIAS - ECOS NORD grant number RC 0899-2012 with the help of ICETEX, and by COLCIENCIAS grant numbers 110656933958 RC 0384-2013 and 123365843539 RC FP44842-081-2014.

Appendix

We will show in this appendix why the $\mathcal{L}_{4,A}^{\rm Gal,bis}$ Lagrangian, uncovered in Ref. [17], is redundant[1].

Let's start with the definitions

$$\mathcal{L}_{4,A}^{\rm Gal,bis} \equiv g_4(A^2)\ A^\mu \tilde{F}_{\mu\nu} S^{\nu\lambda} A_\lambda\,, \tag{A.1}$$

and

$$F_{\mu\nu} \equiv \partial_\mu A_\nu - \partial_\nu A_\mu\,, \tag{A.2}$$

$$S_{\mu\nu} \equiv \partial_\mu A_\nu + \partial_\nu A_\mu\,. \tag{A.3}$$

[1] This is an unpublished proof by Ryo Namba. We thank him for sharing the proof and giving us permission to reproduce it here.

Thus,

$$\begin{aligned}
\mathcal{L}_{4,A}^{\text{Gal,bis}} &\equiv g_4(A^2)\; A^\mu \tilde{F}_{\mu\nu} S^{\nu\lambda} A_\lambda \\
&= g_4(A^2)\; A^\mu A_\lambda \tilde{F}_{\mu\nu} (F^{\lambda\nu} + 2\, \partial^\nu A^\lambda) \\
&= g_4(A^2)\; A^\mu A_\lambda \left(\frac{1}{4} \tilde{F}_{\rho\sigma} F^{\rho\sigma} \delta^\lambda_\mu + 2\tilde{F}_{\mu\nu}\, \partial^\nu A^\lambda \right) \\
&= g_4(A^2) \left(\frac{1}{4} A^2 F \cdot \tilde{F} + A^\mu \tilde{F}_{\mu\nu}\, \partial^\nu A^2 \right) \\
&= \frac{g_4(A^2)}{4} A^2 F \cdot \tilde{F} + g_4(A^2) \partial_\nu A^2 A_\mu \tilde{F}^{\mu\nu} ,
\end{aligned} \tag{A.4}$$

where we have used, in the third line, the identity $\tilde{F}_{\mu\nu} F^{\lambda\nu} = \frac{1}{4} \tilde{F}_{\rho\sigma} F^{\rho\sigma} \delta^\lambda_\mu$ derived for the first time in Ref. [40] and employed in Ref. [17].

Let's now define

$$G_4(A^2) \equiv \int^{A^2} d(A^2)' g_4[(A^2)'] , \tag{A.5}$$

so that

$$\partial_\nu G_4(A^2) = G_{4,A^2} \partial_\nu A^2 = g_4(A^2) \partial_\nu A^2 . \tag{A.6}$$

Therefore,

$$\begin{aligned}
\mathcal{L}_{4,A}^{\text{Gal,bis}} &\equiv \frac{g_4(A^2)}{4} A^2 F \cdot \tilde{F} + \partial_\nu G_4(A^2) A_\mu \tilde{F}^{\mu\nu} \\
&= \frac{g_4(A^2)}{4} A^2 F \cdot \tilde{F} + \partial_\nu \left(G_4(A^2) A_\mu \tilde{F}^{\mu\nu} \right) - G_4(A^2) \partial_\nu (A_\mu \tilde{F}^{\mu\nu}) \\
&= \frac{g_4(A^2)}{4} A^2 F \cdot \tilde{F} + \partial_\nu \left(G_4(A^2) A_\mu \tilde{F}^{\mu\nu} \right) - G_4(A^2) \left(\frac{1}{2} F_{\nu\mu} \tilde{F}^{\mu\nu} + A_\mu \partial_\nu \tilde{F}^{\mu\nu} \right) .
\end{aligned} \tag{A.7}$$

The Bianchi identity $\partial_\alpha F_{\beta\gamma} + \partial_\gamma F_{\alpha\beta} + \partial_\beta F_{\gamma\alpha} = 0$ leads us immediately to conclude that $\tilde{F}^{\mu\nu}$ is divergenceless. The final conclusion is, therefore,

$$\mathcal{L}_{4,A}^{\text{Gal,bis}} \equiv \frac{g_4(A^2) A^2 + 2G_4(A^2)}{4} F \cdot \tilde{F} + \partial_\nu \left(G_4(A^2) A_\mu \tilde{F}^{\mu\nu} \right) , \tag{A.8}$$

where the first term belongs to $\mathcal{L}_{2,A}^{\text{Gal}}$ and the second one may be discarded as it is a total derivative. The curved spacetime generalization of this result still applies due to the Bianchi identity of the Riemann tensor that guarantees that $\tilde{F}^{\mu\nu}$ is still divergenceless.

References

[1] Ostrogradski M 1850 *Mem. Ac. St. Petersbourg* **VI** 385
[2] Woodard R P 2015 *Scholarpedia* **10** 32243 (*Preprint* 1506.02210)
[3] Woodard R P 2007 *Lect. Notes Phys.* **720** 403–433 (*Preprint* astro-ph/0601672)
[4] Horndeski G W 1974 *Int. J. Theor. Phys.* **10** 363–384
[5] Horndeski G W 1976 *J. Math. Phys.* **17** 1980–1987
[6] Deffayet C, Gao X, Steer D and Zahariade G 2011 *Phys. Rev. D* **84** 064039 (*Preprint* 1103.3260)
[7] Kobayashi T, Yamaguchi M and Yokoyama J 2011 *Prog. Theor. Phys.* **126** 511–529 (*Preprint* 1105.5723)
[8] Nicolis A, Rattazzi R and Trincherini E 2009 *Phys. Rev. D* **79** 064036 (*Preprint* 0811.2197)
[9] Deffayet C and Steer D A 2013 *Class. Quant. Grav.* **30** 214006 (*Preprint* 1307.2450)
[10] Maleknejad A, Sheikh-Jabbari M M and Soda J 2013 *Phys. Rept.* **528** 161–261 (*Preprint* 1212.2921)

[11] Dimopoulos K 2012 *Int. J. Mod. Phys. D* **21** 1250023 [Erratum: Int. J. Mod. Phys. D 21,1292003 (2012)] (*Preprint* 1107.2779)
[12] Deffayet C, Gümrükçüoglu A E, Mukohyama S and Wang Y 2014 *JHEP* **1404** 082 (*Preprint* 1312.6690)
[13] Heisenberg L 2014 *JCAP* **1405** 015 (*Preprint* 1402.7026)
[14] Tasinato G 2014 *JHEP* **1404** 067 (*Preprint* 1402.6450)
[15] Allys E, Peter P and Rodríguez Y 2016 *JCAP* **1602** 004 (*Preprint* 1511.03101)
[16] Beltrán Jimenez J and Heisenberg L 2016 *Phys. Lett. B* **757** 405–411 (*Preprint* 1602.03410)
[17] Allys E, Beltrán Almeida J P, Peter P and Rodríguez Y 2016 *JCAP* **1609** 026 (*Preprint* 1605.08355)
[18] Deffayet C, Esposito-Farese G and Vikman A 2009 *Phys. Rev. D* **79** 084003 (*Preprint* 0901.1314)
[19] Deffayet C, Deser S and Esposito-Farese G 2009 *Phys. Rev. D* **80** 064015 (*Preprint* 0906.1967)
[20] Beltrán Jiménez J, Durrer R, Heisenberg L and Thorsrud M 2013 *JCAP* **1310** 064 (*Preprint* 1308.1867)
[21] Rodríguez Y 2015 (*Preprint* 1512.03827)
[22] Allys E 2016 (*Preprint* 1612.01972)
[23] Allys E, Peter P and Rodríguez Y 2016 *Phys. Rev. D* **94** 084041 (*Preprint* 1609.05870)
[24] Padilla A, Saffin P M and Zhou S Y 2011 *Phys. Rev. D* **83** 045009 (*Preprint* 1008.0745)
[25] Metha M L, Normand J M and Gupta V 1983 *Commun. Math. Phys.* **90** 69–78
[26] Beltrán Jiménez J and Heisenberg L 2016 (*Preprint* 1610.08960)
[27] Navarro A A and Rodríguez Y 2017, Work in progress
[28] Rodríguez Y, Gómez L G and Nieto C M 2015 Towards a proof of the equivalence between FRW background expansion and statistical isotropy *Proceedings, 2nd Argentinian-Brazilian Meeting on Gravitation, Relativistic Astrophysics and Cosmology (GRACo II): Buenos Aires, Argentina, April 22-25, 2014* pp 139–144 (*Preprint* 1502.07292)
[29] Kim J and Komatsu E 2013 *Phys. Rev. D* **88** 101301 (*Preprint* 1310.1605)
[30] Ramazanov S R and Rubtsov G 2014 *Phys. Rev. D* **89** 043517 (*Preprint* 1311.3272)
[31] Ade P A R *et al.* (Planck Collaboration) 2016 *Astron. Astrophys.* **594** A16 (*Preprint* 1506.07135)
[32] Ramazanov S, Rubtsov G, Thorsrud M and Urban F R 2016 (*Preprint* 1612.02347)
[33] Maleknejad A and Sheikh-Jabbari M M 2013 *Phys. Lett. B* **723** 224–228 (*Preprint* 1102.1513)
[34] Maleknejad A and Sheikh-Jabbari M M 2011 *Phys. Rev. D* **84** 043515 (*Preprint* 1102.1932)
[35] Nieto C M and Rodríguez Y 2016 *Mod. Phys. Lett. A* **31** 1640005 (*Preprint* 1602.07197)
[36] Adshead P and Wyman M 2012 *Phys. Rev. Lett.* **108** 261302 (*Preprint* 1202.2366)
[37] Davydov E and Galtsov D 2016 *Phys. Lett. B* **753** 622–628 (*Preprint* 1512.02164)
[38] Beltrán Jiménez J, Heisenberg L, Kase R, Namba R and Tsujikawa S 2017 (*Preprint* 1702.01193)
[39] Guarnizo A and Rodríguez Y 2017, Work in progress
[40] Fleury P, Beltrán Almeida J P, Pitrou C and Uzan J P 2014 *JCAP* **1411** 043 (*Preprint* 1406.6254)

IOP Conf. Series: Journal of Physics: Conf. Series **831** (2017) 012005 doi:10.1088/1742-6596/831/1/012005

Black hole geometrothermodynamics

Hernando Quevedo

E-mail: quevedo@nucleares.unam.mx
Instituto de Ciencias Nucleares, Universidad Nacional Autónoma de México, AP 70543, México, DF 04510, Mexico
Dipartimento di Fisica and ICRANet, Università di Roma "La Sapienza", I-00185 Roma, Italy

Abstract.
We review the main aspects of geometrothermodynamics which is a geometric formalism to describe thermodynamic systems, taking into account the invariance of classical thermodynamics with respect to Legendre transformations. We focus on the particular case of black holes, and present a Riemannian metric which describes the corresponding space of equilibrium states. We show that this metric can be used to describe the stability properties and phase transition structure of black holes in different gravity theories.

1. Introduction

The main goal of geometrothermodynamics (GTD) is to construct a formalism that allows us to investigate the properties of thermodynamic systems by using the geometric properties of the corresponding equilibrium spaces [1]. In fact, the idea is that to any given thermodynamic system we can associate an equilibrium space whose points represent possible states at which the system can remain at equilibrium. This means that at any given moment of time the system can be represented geometrically as existing on a particular point of the equilibrium space.

To develop such a geometric formalism, we use as a guiding principle the already existing geometric representation of classical fields. Consider, for instance, the gravitational field. According to general relativity, a gravitational field can be represented by a 4-dimensional Riemannian manifold, the spacetime, whose metric satisfies the Einstein equations. Two geometric concepts are very important in this construction. First, the Riemannian curvature is known to represent the gravitational interaction. This is to say, if no gravitational field is presented, it is expected that the Riemannian curvature vanishes. This is true in general relativity since from the lack of curvature it follows the flat spacetime of special relativity. The second important concept is that of curvature singularities. Intuitively, we expect that singularities indicate the limits of applicability of the theory because they correspond to an infinite gravitational interaction which cannot be treated correctly in general relativity. In fact, the presence of singularities is a major problem in general relativity that is expected to be completely solved only in the framework of a theory of quantum gravity, which is not yet developed.

Other classical fields permit also a geometric representation in terms of principal fiber bundles. In the case of gauge fields, one can take the base space as the flat Minkowski spacetime and the standard fiber as the gauge symmetry. The corresponding curvature of the principal fiber bundle turns out to correspond to the gauge field strength, and the curvature singularities again indicate the breakdown of the underlying classical gauge field theory. In this way, we see that all the classical field interactions existing in Nature permit a representation in terms of concepts of differential geometry [2]. In all the above examples of classical interactions, the explicit construction of the theory is based upon the use of the diffeomorphism and gauge invariance for the gravitational and gauge fields, respectively. These are very important symmetries of the theory that have a very deep physical significance.

Now we ask the question whether it is possible to construct a geometric formalism for the theory of classical equilibrium thermodynamics [3] such that the curvature represents the thermodynamic interaction and the singularities indicate the breakdown of the classical theory. GTD is an attempt to construct such a geometric formalism. An important ingredient of GTD is the Legendre invariance which will play a role similar to that of physical symmetries in classical fields. As mentioned above, the idea is to equip the equilibrium space with a geometric structure which allows one to define the curvature of that space. This can be done in several ways. The formalism of thermodynamic geometry,for instance, consists in introducing different Hessian metrics, computed from different thermodynamic potentials, into the equilibrium space [4, 5, 6]. The interesting feature of using Hessian metrics in the equilibrium space is that their components acquire a very interesting physical interpretation in fluctuation theory. In fact, small fluctuations of the thermodynamic potential can be decomposed into moments, and it turns out that the second moment of the fluctuation is essentially determined by the components of the corresponding Hessian metric. Nevertheless, some difficulties have been found when applying Hessian metrics to certain thermodynamic systems. It turns out that different Hessian metrics predict different properties for the same system. In other words, the properties of a thermodynamic system, as represented in thermodynamic geometry, depend on the choice of thermodynamic potential. This is not consistent with the results of classical thermodynamics [3]. The formalism of GTD avoids these difficulties by using only geometric objects which are invariant with respect to Legendre transformations.

In this work, we review the main aspects of the GTD formalism and apply it to characterize the thermodynamic properties of black holes from a geometric point of view. In Sec. 2, we discuss the properties of Legendre transformations when interpreted as transformations between the coordinates of an abstract space, the phase space, which is explicitly constructed in Sec. 3. We show that the phase space can be endowed with a contact structure and a Riemannian structure, both being invariant with respect to Legendre transformations. The definition of the equilibrium manifold is given in Sec. 4 as a submanifold of the phase manifold which inherits its invariance properties. Then, in Sec. 5, we investigate explicitly the geometrothermodynamic properties of black holes in Einstein's gravity. Finally, in Sec. 6, we discuss our results and some tasks of future research. We use throughout geometric units with $c = G = \hbar = k_B = 1$.

2. Legendre invariance

In classical equilibrium thermodynamics, to describe a system with n degrees of freedom, it is necessary to specify n extensive variables E^a $(a = 1, ..., n)$, their intensive duals I^a and a thermodynamic potential Φ. Let us for the moment suppose that all these variables are independent from each other, and denote them by $Z^A = \{\Phi, E^a, I^a\}$. We can therefore use this set as coordinates of an abstract space. If, in addition, we suppose that this space is differentiable, we can use the standard procedures of differential geometry in order to introduce other coordinates by means of diffeomorphisms. In particular, we can consider Legendre transformations as coordinate transformations in this abstract space and investigate their properties from the point of view of differential geometry. This is one of the main assumptions of GTD, and we will see that it leads to consistent results. In addition, this opens the possibility to consider general diffeomorphisms that act on the geometric structure of GTD.

Let us now define a total Legendre transformation[1] by means of [7]

$$Z^A \to \tilde{Z}^A = \{\tilde{\Phi}, \tilde{E}^a, \tilde{I}^a\} \tag{1}$$

with

$$\Phi = \tilde{\Phi} - \tilde{E}^a \tilde{I}_a \ , \quad E^a = -\tilde{I}^a, \quad I^a = \tilde{E}^a \ . \tag{2}$$

This representation is equivalent to the standard one which is used in the equilibrium space. However, it allows us to consider a Legendre transformation as a coordinate transformation determined by the matrix

$$\frac{\partial Z^A}{\partial \tilde{Z}^B} = \begin{pmatrix} 1 & -\tilde{Z}^{n+1} & -\tilde{Z}^{n+2} & \dots & -\tilde{Z}^{2n} & -\tilde{Z}^1 & -\tilde{Z}^2 & \dots & -\tilde{Z}^n \\ 0 & 0 & 0 & \dots & 0 & -1 & 0 & \dots & 0 \\ . & . & . & \dots & . & . & . & \dots & . \\ 0 & 0 & 0 & \dots & 0 & 0 & 0 & \dots & -1 \\ 0 & 1 & 0 & \dots & 0 & 0 & 0 & \dots & 0 \\ . & . & . & \dots & . & . & . & \dots & . \\ 0 & 0 & 0 & \dots & 1 & 0 & 0 & \dots & 0 \end{pmatrix} , \tag{3}$$

where the indices A and B represent rows and columns, respectively. The determinant of this matrix is equal to one. Therefore, one can compute the inverse transformation which is the condition for the above transformation to be considered as a diffeomorphism.

As we see from the above expressions, a Legendre transformation interchanges the role of the extensive and intensive variables. In this sense, it resembles the role of canonical transformations of classical mechanics in which the generalized coordinates can become momenta by means of a canonical transformation.

Our goal is to construct geometric structures which are invariant with respect to Legendre transformations. This is similar to the Lorentz invariance of special relativity. Indeed, consider, for instance, the 2−dimensional Minkowski metric in Cartesian coordinates

$$ds^2 = -dt^2 + dx^2 \ . \tag{4}$$

If we apply a special Lorentz transformation

$$(t, x) \to (\tilde{t}, \tilde{x}) \tag{5}$$

given by the relationships

$$t = \frac{1}{\sqrt{1-v^2}} \left(\tilde{t} + v\tilde{x}\right) \ , \quad x = \frac{1}{\sqrt{1-v^2}} \left(\tilde{x} + v\tilde{t}\right) \ , \tag{6}$$

where v is a constant, the line element (4) transforms into

$$ds^2 = -d\tilde{t}^2 + d\tilde{x}^2 \ , \tag{7}$$

which has the same functional dependence as the original line element.

Let us now consider the abstract space with coordinates Z^A. As an example, we check if the flat metric

$$ds^2 = \delta_{AB} dZ^A dZ^B \tag{8}$$

[1] Partial Legendre transformations can be defined analogously. However, in this work, we limit ourselves to the study of total transformations.

is Legendre invariant. For concreteness, consider the case $n = 1$, i.e.,

$$ds^2 = d\Phi^2 + (dE^1)^2 + (dI^1)^2 \,. \tag{9}$$

Applying the transformation (2), it is straightforward to see that the flat metric transforms into

$$ds^2 = d\tilde{\Phi}^2 - 2d\tilde{\Phi}(\tilde{E}^1 d\tilde{I}^1 + \tilde{I}^1 d\tilde{E}^1) + 2\tilde{E}^1\tilde{I}^1 d\tilde{E}^1 d\tilde{I}^1 + (d\tilde{E}^1)^2 + (d\tilde{I}^1)^2 \,, \tag{10}$$

whose functional dependence is clearly different. We conclude that a Legendre invariant metric is necessarily curved.

In general, it is not easy to find Legendre invariant quantities because the set of Legendre transformations do not form a group. Moreover, in the case of subsets which form a group, one can show that not all them can be obtained from the generators of infinitesimal Legendre transformations [8]. The search for Legendre invariant geometric quantities must therefore be conducted in an empiric way in which the functional invariance should be represented explicitly, giving rise to a set of algebraic equations to be solved by using the standard methods of linear algebra. This is the method we will use in the following sections.

3. Phase manifold

One the main geometric components of GTD is the phase manifold which is necessary in order to represent Legendre transformations as diffeomorphisms. The triad $(\mathcal{T}, \Theta, G)$ is called phase manifold if the following conditions are satisfied.

(i) $\mathcal{T}$ is a $(2n+1)-$dimensional differential manifold, a local patch of which is characterized by the set of coordinates $Z^A = \{\Phi, E^a, I^a\}$.

(ii) Θ is a contact 1−form defined on the cotangent manifold $T^*_{\mathcal{T}}$, i.e., it satisfies the condition $\Theta \wedge (d\Theta)^n \neq 0$, meaning that $\mathcal{T}$ is maximally non-integrable. According to Darboux theorem, locally it is possible to express the contact 1−form as

$$\Theta = d\Phi - I_a dE^a \,. \tag{11}$$

(iii) G is a Legendre invariant Riemannian metric on $\mathcal{T}$, i.e., $G = G_{AB} dZ^A dZ^B$.

In general, the triad $(\mathcal{T}, \Theta, G)$ represents a Riemannian contact manifold. The contact form is Legendre invariant in the sense that under a Legendre transformation $Z^A \to \tilde{Z}^A$, it behaves as

$$\Theta \to \tilde{\Theta} = d\tilde{\Phi} - \tilde{I}_a d\tilde{E}^a \,. \tag{12}$$

In GTD, we also demand that the metric G be Legendre invariant. As mentioned in the previous section, it is difficult to generate Legendre invariant objects due to the non-standard properties of the set of Legendre transformations. Therefore, in order to construct a Legendre invariant metric, we proceed as follows. We start from a general metric $G = G_{AB} dZ^A dZ^B$ and transform the components G_{AB} with the transformation matrix (3). For the resulting components $\tilde{G}_{AB}$ we demand functional invariance as explained above. This leads to a system of algebraic equations that can be solved to obtain specific Legendre invariant metrics [1]. Among the particular solutions we have found so far [9, 10], the metric

$$G^{II} = \Theta^2 + (\xi_{ab} I^a E^b)(\eta_{cd} dI^c dE^d) \,, \tag{13}$$

has been shown to be applicable in the case of thermodynamic systems with second-order phase transitions, like black holes. Here $\eta_{cd} = \mathrm{diag}(-1, \cdots, 1)$ and ξ_{ab} is a diagonal constant matrix.

It is important to note that to the above matrix structure it is still possible to add quadratic terms [1, 8]

$$\delta_{ab}(dE^a dE^b + dI^a dI^b) \ , \tag{14}$$

non-conformal terms

$$\delta_{ab} E^b I^a dE^a dI^b \ , \tag{15}$$

and mixed conformal terms

$$\Omega\, \epsilon_{ab} dE^a dI^b \ , \tag{16}$$

where $\delta_{ab} = \mathrm{diag}(1, \cdots, 1)$, $\epsilon_a^{\ b} = 1, -1, 0$ for $a > b$, $a < b$, $a = 0$, respectively, and Ω is an arbitrary Legendre invariant function of the coordinates E^a and I^a. However, as we will see below, none of them satisfies the physical condition that the thermodynamic curvature of the equilibrium manifold vanishes when thermodynamic interaction is lacking. In this sense, the metric (13) is essentially unique. In fact, the contact form Θ can still be multiplied by an arbitrary non-zero function of the coordinates Z^A which, however, does not affect the geometric structure of the physical equilibrium manifold. Moreover, the second term of the metric (13) is determined modulo a multiplicative constant which can be set equal to one, without loss of generality.

We thus see that the phase manifold of GTD is a Legendre invariant, Riemannian, contact manifold [11]. The above procedure shows that the construction of the phase manifold is essentially unique. Indeed, once the coordinates of $\mathcal{T}$ are fixed, the contact form is given in a canonical way, modulo a non-zero conformal function which does not affect the physical results. The metric structure G is also essentially unique as explained above. The constant diagonal matrix ξ_{ab} which enters the metric (13) contains, in principle, n arbitrary constants which are not fixed by the Legendre invariance condition. However, they can be fixed by imposing an additional physical condition which is related to the Euler identity as we will see in the following sections.

4. Equilibrium manifold

In the previous section, we introduced the concept of phase manifold as an auxiliary space to handle Legendre transformations as diffeomorphisms. On the other hand, we mentioned that thermodynamic geometry uses only the equilibrium space to describe thermodynamic systems from a geometric perspective, without considering the Legendre invariance of classical equilibrium thermodynamics, i.e., without considering the corresponding phase space. In this section, we will show that in fact it is possible to consider the equilibrium space of any thermodynamic system as subspace of the phase manifold.

In GTD, an equilibrium manifold is a Riemannian manifold $(\mathcal{E}, g)$ defined by the following conditions:

(i) $\mathcal{E} \in \mathcal{T}$ with $\dim(\mathcal{E}) = n$.

(ii) There exists a smooth embedding map $\varphi : \mathcal{E} \to \mathcal{T}$ such that

$$\varphi^*(\Theta) = 0 \ , \tag{17}$$

where φ^*,is the pullback of φ.

(iii) There exists a canonically induced metric g determined by

$$g = \varphi^*(G) \ . \tag{18}$$

IOP Conf. Series: Journal of Physics: Conf. Series **831** (2017) 012005 doi:10.1088/1742-6596/831/1/012005

We see that the definition of an equilibrium manifold is closely related with the existence of the smooth map φ. It is therefore important to establish the conditions under which the map φ can exist. To this end, consider the explicit form of the map φ in a particular coordinate system. Let us consider E^a as the coordinates of $\mathcal{E}$. Then, the map $\varphi : \mathcal{E} \to \mathcal{T}$ implies that

$$\varphi : \{Z^A\} \mapsto \{\Phi(E^a), E^a, I^a(E^a)\} , \tag{19}$$

i.e, all the coordinates of $\mathcal{T}$ should now depend on the coordinates of $\mathcal{E}$. Consider the function $\Phi(E^a)$. As mentioned in the previous section, Φ was introduced as a thermodynamic potential and E^a are the extensive variables. Then, $\Phi(E^a)$ represents a fundamental thermodynamic equation and Φ can be either the entropy or the internal energy of the system [3]. We conclude that the equilibrium manifold can be defined only if the fundamental equation is known. This is in accordance with the results of classical thermodynamics in which all the properties of a system can be obtained from the explicit form of the fundamental equation. Now, we have in GTD a similar situation. We need to know the fundamental equation $\Phi(E^a)$, as part of the smooth map φ, in order to construct the equilibrium manifold. This implies that for a given thermodynamic system, given by a unique fundamental equation $\Phi(E^a)$, there exists a unique equilibrium manifold $\mathcal{E}$.

Consider now condition (17) which in coordinates Z^A implies that

$$d\Phi = I_a dE^a \ , \quad I_a = \frac{\partial \Phi}{\partial E^a} \ . \tag{20}$$

This implies that $I_a = I_a(E^a)$, as demanded by the map (19), and that I_a is the intensive variable dual to E^a as in classical thermodynamics. On the other hand, we can now interpret the condition $\varphi^*(\Theta) = 0$ as the first law of thermodynamics.

The condition (18) determines the metric of the equilibrium manifold which in the case of the Legendre invariant metric (13) can be expressed as

$$g^{^{II}} = \varphi^*(G^{^{II}}) = \left(\xi_a^{\ b} E^a \Phi_{,b}\right) \left(\eta_a^b \Phi_{,bc} \, dE^a dE^c\right) \ , \tag{21}$$

where $\eta_a^b = \text{diag}(-1, 1, \cdots, 1)$, $\xi_a^{\ b} = \xi_{ac}\delta^{bc}$, and $\Phi_{,a} = \frac{\partial \Phi}{\partial E^a}$. We see that all the components of the metric $g^{^{II}}$ can be calculated explicitly once the fundamental equation $\Phi(E^a)$ is given. The conformal term can be put proportional to the potential Φ by using Euler's identity. Indeed, if Φ is a homogeneous function of degree β, i.e., $\Phi(\lambda E^a) = \lambda^\beta \Phi(E^a)$, Euler's identity reads $E^a \Phi_{,a} = \beta \Phi$. If Φ is instead a generalized homogeneous function [18], i.e., $\Phi(\lambda^{\alpha_a} E^a) = \lambda^{\alpha_\Phi} \Phi(E^a)$, then $\alpha_a E^a \Phi_{,a} = \alpha_\Phi \Phi$. Therefore, we see that it is always possible to choose the components of the diagonal matrix $\xi_a^{\ b}$ proportional to α_a so that the conformal factor becomes proportional to Φ, i.e.,

$$g^{^{II}} = \Phi \left(\eta_a^b \Phi_{,bc} \, dE^a dE^c\right) \ . \tag{22}$$

We conclude that the equilibrium manifold as defined in GTD contains in a canonical manner the information about the first law of thermodynamics and the fundamental equation of the thermodynamical system under consideration. The second law of thermodynamics should be contained in the fundamental equation as a consequence of the condition that the corresponding system is physical. The third law implies that there exists a point in the equilibrium manifold which cannot be reached by a finite number of quasi-static processes. This implies geometrically that the point of minimum entropy should not be part of the equilibrium manifold. In the case of the ideal gas, this implies that the topology of the equilibrium manifold is not trivial [12].

In the following diagram, we summarize the geometric structure of GTD.

$$
\begin{array}{ccc}
\mathcal{E} & \xrightarrow{\varphi} & \mathcal{T} \\
\downarrow & & \downarrow \\
T_{\mathcal{E}} & \xrightarrow{\varphi_*} & T_{\mathcal{T}} \\
\downarrow & & \downarrow \\
T^*_{\mathcal{E}} & \xleftarrow{\varphi^*} & T^*_{\mathcal{T}} \\
g & \xleftarrow{\varphi^*} & G
\end{array}
$$

Due to the differentiability properties of the phase manifold $\mathcal{T}$, it is possible to introduce the corresponding tangent $T_{\mathcal{T}}$ and cotangent space $T^*_{\mathcal{T}}$. Since the embedding map φ is smooth, it is also possible to define in a canonical way the pushforward φ_* and pullback φ^* that act on the tangent and cotangent spaces, respectively. Finally, once the metric G is given on $\mathcal{T}$, the embedding map induces a canonical metric g on $\mathcal{E}$.

It is important to mention that if we insert the fundamental equation for an ideal gas

$$S = \frac{3}{2}\ln U + \ln V \ , \tag{23}$$

where U is the internal energy, V is the volume and $N = 1$ is the molar number, the curvature of the metric (22) vanishes, implying that no thermodynamic interaction exists, as expected from classical thermodynamics. However, if we add anyone of the Legendre invariant terms (14), (15) or (16), we obtain a non-flat equilibrium manifold. Because of this inconsistency, those additional terms are not considered in GTD.

5. Black holes

Since the discovery that the area of a black hole horizon satisfies a series of rules that resemble the laws of thermodynamics [13, 14, 15], black holes have been intensively studied by using the standard methods of classical thermodynamics [16]. It is in this sense that black holes can be considered as thermodynamic systems. However, this is still a controversial approach in particular because a definite microscopic model for black holes is still missing [17].

The main postulate of black hole thermodynamics is that the entropy S is proportional to the horizon area A as

$$S = \frac{1}{4}A \ , \tag{24}$$

which can be considered as the fundamental equation of black holes. This implies that the entropy is a function of the extensive variables $S = S(E^a)$ which correspond to the physical parameters of the black hole.

Consider now the most general black hole in Einstein's theory whose spacetime is completely determined by the Kerr metric [19]

$$
\begin{aligned}
ds^2 = & - \frac{\Delta - a^2\sin^2\theta}{\Sigma}dt^2 - \frac{2a\sin^2\theta(r^2+a^2-\Delta)}{\Sigma}dtd\varphi \\
& + \frac{(r^2+a^2)^2 - a^2\sin^2\theta\,\Delta}{\Sigma}\sin^2\theta d\varphi^2 + \frac{\Sigma}{\Delta}dr^2 + \Sigma d\theta^2 \ ,
\end{aligned} \tag{25}
$$

$$\Sigma = r^2 + a^2 \cos^2 \theta \ , \quad \Delta = (r - r_+)(r - r_-) \ , \quad r_\pm = M \pm \sqrt{M^2 - a^2} \ , \tag{26}$$

where M is the total mass of the black hole and $a = J/M$ is the specific angular momentum. It is then straighforward to compute the area of the exterior horizon to obtain

$$S = 2\pi(M^2 + \sqrt{M^4 - J^2}) \ . \tag{27}$$

To construct the GTD of the Kerr black hole, we first have to identify the thermodynamic variables. Since $S = \Phi$ is the thermodynamic potential, then $E^a = (M, J)$. Therefore, the metric of the equilibrium manifold (22) becomes

$$g^{^K} = S\left(-S_{MM}dM^2 + S_{JJ}dJ^2\right) \ , \tag{28}$$

where the subscript denotes partial derivative. Now, the corresponding scalar curvature can be shown to be of the form

$$R^{^K} = \frac{N^{^K}}{S_{MM}S_{JJ}} \ , \tag{29}$$

where $N^{^K}$ is a function of M and J which is in general non-zero. The fact that the curvature of the the equilibrium manifold is non-vanishing indicates that the Kerr black hole is a system with thermodynamic interaction. Moreover, we see that curvature singularities appear when $S_{MM} = 0$ and $S_{JJ} = 0$ which, according to the interpretation of the thermodynamic curvature, implies the existence of phase transitions. On the other hand, if we interpret black holes as classical thermodynamic systems, it is possible to explore the behavior of the response functions in order to find the phase transition structure. Indeed, according to the Ehrenfest scheme [3], a zero or a discontinuity in the $n-$th derivative of a thermodynamic potential indicates the presence of an $n-$th phase transition. We conclude that the thermodynamic curvature of the equilibrium manifold of the Kerr black hole predicts the existence of second order phase transitions. The location of the transition points in the equilibrium space has been derived explicitly in [20].

We now consider the Einstein-Maxwell theory. The most general black hole in this case is described by the Kerr-Newman metric which is again given by Eq.(25) with different values for the horizon radii, namely,

$$r_\pm = M \pm \sqrt{M^2 - a^2 - Q^2} \ , \tag{30}$$

where Q is the electric charge of the black hole. Then, the fundamental equation is given by

$$S = \pi\left(2M^2 - Q^2 + 2\sqrt{M^4 - J^2 - M^2Q^2}\right) \ , \tag{31}$$

indicating that the coordinates of the equilibrium manifold are $E^a = (M, J, Q)$. The GTD of the Kerr-Newman black hole is then determined by the thermodynamic metric

$$g^{^{KN}} = S\left(-S_{MM}dM^2 + S_{JJ}dJ^2 + S_{QQ}dQ^2 + 2S_{JQ}dJdQ\right) \ . \tag{32}$$

We notice that in this case there is some ambiguity in the choice of the coordinate which should be associated with the minus sign of the matrix $\eta_a^b = \text{diag}(-1, 1, 1)$. Since there are three different possible choices, it turns out that the right choice corresponds to the coordinate M. The reason is that the minus sign should be associated with the coordinates which determine the thermodynamic potential of the fundamental equation, i.e, the entropy or the internal energy. This shows that although in GTD it is possible to use any thermodynamic potential which is obtained from the entropy or the internal energy by means of a total Legendre transformation, the canonical variables associated with the fundamental equation are in certain sense privileged.

A straightforward computation of the curvature tensor of the metric (32) indicates that in general several components are different from zero. This means that the Kerr-Newman black hole is characterized by thermodynamic interaction. For the study of the curvature singularities, we have to investigate scalars formed by the components of the curvature tensor. The Ricci scalar $R = g^{ab}g^{cd}R_{acbd}$ is the simplest scalar available, but in the 3−dimensional case under consideration, there are other independent scalars, in particular, those which are quadratic in the curvature tensor, namely, the Kretschmann scalar $R_{abcd}R^{abcd}$, the Euler scalar $*R_{abcd}R^{abcd}$, and the Pontrjagin scalar $*R_{abcd} * R^{abcd}$, where $*R_{abcd}$ is the dual curvature tensor [21]. A computation of all these scalars shows that all of them yield the same singularities, which are equivalent to the singularities of the Ricci scalar, i.e.,

$$R^{KN} = \frac{N^{KN}}{S_{MM}(S_{JJ}S_{QQ} - S_{JQ}^2)} , \tag{33}$$

where N^{KN} is a non-zero function of the coordinates M, J and Q. We see that there two types of singularities for

$$S_{MM} = 0 , \quad \text{and} \quad S_{JJ}S_{QQ} - S_{JQ}^2 = 0 . \tag{34}$$

The first type of singularities corresponds again to a second order phase transition as can be corroborated by calculating the corresponding response function [20]. The second type of singularities is connected with

$$S_{JJ}S_{QQ} - S_{JQ}^2 \geq 0 , \tag{35}$$

which corresponds to the convexity and stability conditions for any thermodynamic system with J and Q as thermodynamic degrees of freedom [3]. It follows that the second type of singularities implies that during the phase transition the black hole passes from an unstable state ($S_{JJ}S_{QQ} - S_{JQ}^2 < 0$) to a stable state ($S_{JJ}S_{QQ} - S_{JQ}^2 > 0$).

These results show that in fact the curvature of the equilibrium manifold contains the thermodynamic information of black holes.

6. Final remarks

In this work, we reviewed the main geometric structures which are necessary in GTD to develop a formalism which is independent with respect to Legendre transformations. The formalism involves concepts of contact geometry, Riemannian geometry and classical equilibrium thermodynamics. Here, we focused of the applications of GTD in the context of black hole thermodynamics.

First, we introduce the concept of phase manifold in order to handle Legendre transformations as coordinate transformations. Since the phase manifold is necessarily odd dimensional, it is possible to introduce in a canonical way a contact structure which generates the first law of thermodynamics. We endow the phase manifold with a Riemannian metric which is invariant with respect to total Legendre transformations. The auxiliary phase manifold contains only geometric objects which are Legendre invariant and so it represents the main ingredient of GTD.

The equilibrium manifold is defined as contained in the phase manifold and its properties are determined by means of an embedding map whose definition contains a fundamental thermodynamic equation. This means that in principle there exists a unique equilibrium manifold for each thermodynamic system. The goal of GTD is to describe the thermodynamic properties of a given system in terms of the geometric properties of the corresponding equilibrium manifold. In particular, we associate curvature with thermodynamic interaction and singularities with phase transitions.

We present as particular example the GTD of black holes and analyze the most general black holes in Einstein and Einstein-Maxwell theories. In the case of the Kerr and Kerr-Newman black holes, the curvature of the corresponding equilibrium manifolds turns out to be non-zero, pointing to the existence of thermodynamic interaction in these black holes. In addition, the curvature singularities of the equilibrium manifolds are shown to correspond to divergences of the response functions, as defined in classical thermodynamics. Moreover, the thermodynamic curvature also contains information about the stability properties of black holes. Indeed, we found that during a phase transition the black hole passes from an unstable state to a stable state.

Finally, we mention that black hole GTD has been also applied to a large number of black hole configurations in different gravity theories like Einstein-Born-Infeld, Einstein-Yang-Mills, etc. (see Ref. [10] and references cited therein). So far, black hole GTD has shown to lead to consistent results in all the cases in which it has been applied. Therefore, GTD represents an alternative method to study thermodynamic properties of black holes.

A particular contribution of GTD to the physics of black holes is the definition of thermodynamic interaction in terms of curvature. This is certainly an invariant method to explore internal properties of black holes. Indeed, in classical black hole thermodynamics there is no method to define interaction. This is only possible at the level of statistical physics in which the presence of a potential term in the Hamiltonian function indicates the existence of thermodynamic interaction. In black hole physics, however, no definite Hamiltonian is known that could yield from statistical physics the fundamental equation in the corresponding thermodynamic limit. We therefore believe that by investigating the behavior of the thermodynamic curvature, it would be possible to extract information about the interaction inside a black hole.

Acknowledgements

It is an honor to dedicate this work to Prof. Luis Herrera and Prof. Rodolfo Gambini, Two great scientists who are dedicating their lives to developing theoretical physics in Latin America, and two great teachers who are an example to follow for all of us. I thank A. Sánchez for helpful discussions. This work has been supported by the UNAM-DGAPA-PAPIIT, Grant No. IN111617.

References

[1] H. Quevedo, J. Math. Phys. **48**, 013506 (2007).
[2] T. Frankel, *The Geometry of Physics: An Introduction* (Cambridge University Press, Cambridge, UK, 1997).
[3] H.B. Callen, *Thermodynamics* (John Wiley & Sons, Inc., New York, 1981).
[4] S. Amari, *Differential-Geometrical Methods in Statistics* (Springer-Verlag, Berlin, 1985).
[5] F. Weinhold, *Classical and Geometrical Theory of Chemical and Phase Thermodynamics* (Wiley, Hoboken, New Jersey, 2009).
[6] G. Ruppeiner, Springer Proc. Phys. **153**, 179 (2014).
[7] V. I. Arnold, *Mathematical Methods of Classical Mechanics* (Springer Verlag, New York, 1980).
[8] D. Garcia and C.S. Lopez-Monsalvo, J. Math. Phys. **55**, 083515 (2014).
[9] H. Quevedo and M. N. Quevedo, Electr. J. Theor. Phys., **2011**, 1 (2011).
[10] H. Quevedo, M. N. Quevedo and A. Sánchez, Phys. Rev. D **94**, 024057 (2016).
[11] R. Hermann, *Geometry, physics and systems* (Marcel Dekker, New York, 1973).
[12] H. Quevedo, A. Sánchez, and A. Vázquez, Gen. Rel. Grav. **47**, 36 (2015);
[13] J. D. Bekenstein, Phys. Rev. D **7**, 2333 (1973).
[14] S. Hawking, Nature 248, 30 (1974).
[15] S. Hawking, Commun. Math. Phys. **43**, 199 (1975).
[16] P. C. W. Davies, Rep. Prog. Phys. **41**, 1313 (1977).
[17] S. Carlip, Int. J. Mod. Phys. D bf 23, 1430023 (2014).
[18] H. E. Stanley, *Introduction to phase transitions and critical phenomena* (Oxford University Press, New York, 1971).

[19] H. Stephani, D. Kramer, M. MacCallum, C. Hoenselaers, and E. Herlt, *Exact solutions of Einstein's field equations* (Cambridge University Press, Cambridge, UK, 2003).
[20] J. L. Álvarez, H. Quevedo, and A. Sánchez, Phys. Rev. D **77**, 084004 (2008).
[21] C. Cherubini, D. Bini, S. Capozziello, R. Ruffini, Int. J. Mod. Phys. D **11**, 827 (2002).

IOP Conf. Series: Journal of Physics: Conf. Series **831** (2017) 012006 doi:10.1088/1742-6596/831/1/012006

Putting gravity in control

C S Lopez-Monsalvo

E-mail: cslopezmo@conacyt.mx
Conacyt-Universidad Autónoma Metropolitana Azcapotzalco
Avenida San Pablo Xalpa 180, Azcapotzalco, Reynosa Tamaulipas, 02200 Ciudad de México, México

I Lopez-Garcia, F Beltran-Carbajal and R Escarela-Perez

Universidad Autónoma Metropolitana Azcapotzalco
Avenida San Pablo Xalpa 180, Azcapotzalco, Reynosa Tamaulipas, 02200 Ciudad de México, México

Abstract. The aim of the present manuscript is to present a novel proposal in Geometric Control Theory inspired in the principles of General Relativity and energy-shaping control.

It has been a pleasure to prepare this contribution to celebrate the two parallel lives of effort and inspiration in Theoretical Physics of Rodolfo Gambini and Luis Herrera. Albeit different in scope, their research shares the common thread of gravity and its geometric principles, together with their ultimate consequences ranging from the very nature of space and time to the astrophysical implications of thermodynamics and gravity. It is thus our wish to take this opportunity to present a different perspective on the use of the same guiding principles that led to General Relativity and explore, even if very briefly, its implications within the area of control theory.

General relativity brought a new paradigm in the understanding of physical reality, establishing a deep connection between geometry and physics through the falsification of the motion of a test body due to a *universal force* [1] by free motion in a curved manifold. In such case, curvature becomes morally tantamount to a universal force field strength. This insight gave rise to a successful geometrization programme for field theories. However, due to the non-universal nature of the other known interactions, the geometrization is slightly different from that of gravity.

In geometric field theories, problems are of two kinds: given the sources determine the curvature, or, given the curvature determine the trajectories of test particles. In this two type of problems, it is the source which determines the background geometry for the motion. Thus, in general, changing the source distribution changes the curvature of the spacetime manifold, and thus the motion of the test particles. Therefore, in principle, one could be able to reproduce any observed motion in space by means of a suitable distribution of energy. By placing appropriate sources here and there, one can *control* the motion of test particles. Of course, one cannot simply engineer and energy-momentum tensor so that test particles follow our desired trajectories. Most of such energy distributions are indeed unphysical from the spacetime point of view. However, this very principle might be applied to a different physical setting, one for which the background

IOP Conf. Series: Journal of Physics: Conf. Series **831** (2017) 012006 doi:10.1088/1742-6596/831/1/012006

geometry is not the spacetime manifold. Such is the case of the geometrization of classical mechanics and the control theory that can be built from it.

We begin by briefly revisiting the geometrization programme of classical mechanics. Let us consider a mechanical systems characterized by n degrees of freedom (DoF) and defined by a Lagrangian function $L : T\mathcal{Q} \longrightarrow \mathbb{R}$. Here, the configuration space is an n dimensional manifold $\mathcal{Q}$ whose tangent bundle is denoted by $T\mathcal{Q}$. The generic problem in geometric mechanics can be stated as follows. Given a pair of points $p_1, p_2 \in \mathcal{Q}$, find the curve $\gamma \subset \mathcal{Q}$

$$\gamma : [0,1] \longrightarrow \mathcal{Q}, \quad \gamma(0) = p_1, \; \gamma(1) = p_2, \tag{1}$$

for which the functional

$$S[\gamma] = \int_0^1 L[\tilde{\gamma}(\tau)] \, \mathrm{d}\tau, \tag{2}$$

is an extremum. Here, $\tilde{\gamma}$ denotes canonical lift of γ to $T\mathcal{Q}$.

$$\tilde{\gamma} : [0,1] \longrightarrow T\mathcal{Q}, \quad \tilde{\gamma} : \tau \longrightarrow [\gamma(\tau), \dot{\gamma}(\tau)] \,. \tag{3}$$

The fact that the *natural* evolution of a mechanical system connecting two given points of the configuration space $\mathcal{Q}$ follows precisely such a path is known as *Hamilton's Principle.* Thus, natural motions solve the Euler-Lagrange equations

$$\mathcal{E}(L) = 0, \tag{4}$$

where $\mathcal{E}$ is the Euler operator [2].

We will consider systems generated by Lagrangian functions of the form

$$L : T\mathcal{Q} \longrightarrow \mathbb{R}, \quad L = T - V, \tag{5}$$

where T and V represent the kinetic and potential energies, respectively. Moreover, we will restrict our analysis to the case where the kinetic energy is defined in terms of a symmetric, non-degenerate and (usually) positive definite second rank tensor field

$$M : T\mathcal{Q} \times T\mathcal{Q} \longrightarrow \mathbb{R}. \tag{6}$$

Namely, those for which the kinetic energy at a given point is written as

$$T|_p = \frac{1}{2} M\left(U_p, U_p\right), \tag{7}$$

where

$$U_p = \left.\frac{\mathrm{d}\gamma}{\mathrm{d}\tau}\right|_{\tau_0} \in T_p\mathcal{Q}, \quad \text{with} \quad \gamma(\tau_0) = p \in \mathcal{Q} \tag{8}$$

is the velocity of the trajectory at the point p. Such systems are called *natural* (cf. [3])

The tensor field (6) satisfies at each point of $\mathcal{Q}$ the properties of an inner product, promoting the configuration space into a Riemannian manifold. Thus, one could try to relate the curves extremizing the action functional S to geodesics of M in $\mathcal{Q}$. However, they only coincide in the case of *free motion*, that is, when there is no potential energy nor external forces. Nonetheless, this observation gives rise to the problem of finding a metric tensor G for the configuration space $\mathcal{Q}$ such that the extremal curves of the action S coincide with the geodesics of G for a natural system defined by Lagrangian function with potential energy function V.

IOP Conf. Series: Journal of Physics: Conf. Series **831** (2017) 012006 doi:10.1088/1742-6596/831/1/012006

Recalling that geodesics are themselves extrema of the arc-length functional of $\mathcal{Q}$, it is a straightforward exercise showing that (c.f. Chapter 4 in [3]), for conservative systems, the metric we are looking for is conformal to M. That is, the solutions to the geodesic equation

$$\nabla_U U = 0, \tag{9}$$

where ∇ is a connection compatible with the metric

$$G = 2\left[E - V(p)\right] M, \tag{10}$$

extremize the action functional (2). Here, E is the energy of the initial conditions and, by assumption, it is a constant of the motion. The metric (10) is called the *Jacobi metric.* There is a difference, however, in the geometric origin of these curves. On the one hand, they are the paths followed by the system in a potential whilst, on the other, those are the free paths of the purely kinetic Lagrangian

$$L_G = \frac{1}{2} G(U, U). \tag{11}$$

Now that we can identify the trajectories in configuration space of the *natural motion* of a given system with geodesics in a Riemannian manifold, we would like to bend those paths so that the system evolves in a *desired* manner, that is, we want to re-shape the geometry so that the *desired evolution* corresponds to geodesic motion in a control Riemannian manifold. Thus, albeit both, classical mechanics and control theory are based on the same dynamical principles, they do differ in their objectives and goals. On the one hand, the generic problem of classical mechanics is that of finding the integral curves to a given Lagrangian vector field whilst, on the other hand, in control theory one is interested in finding the control *inputs*, generated by properly located actuators, so that the integral curves of a given system follow a designed or desired path in configuration space. Moreover, in both cases, the problem of stability is of paramount relevance. In the following lines we present a proposal to address the stability problem in control theory from a Riemannian point of view.

The way one controls the evolution of a system is by directly acting upon a set of *accesible* degrees of freedom (ADoF) [4]

$$\mathcal{D} = \left\{\hat{e}_{(i)}(p)\right\}_{i=1}^{m}, \quad \text{where} \quad \hat{e}_{(i)}(p) \in T_p\mathcal{Q} \quad \forall p \in \mathcal{Q}, \tag{12}$$

is the ith element of a frame over the configuration space so that we can place the control action directly into the Euler-Lagrange equations as

$$\mathcal{E}(L) = M^\sharp\left[u\right] \in T\mathcal{Q}, \quad u = \sum_{i=1}^{m} u_i \hat{f}^{(i)} \quad \text{with} \quad \hat{f}^{(i)} = M^\flat\left[\hat{e}_{(i)}\right], \tag{13}$$

where u represents our control input as an applied external force acting on $\mathcal{D}$. Here, $M^\sharp$ and $M^\flat$ denote the musical isomorphisms between the tangent and co-tangent bundles defined by the kinetic energy metric M, i.e.

$$M^\sharp : T_p^*\mathcal{Q} \longrightarrow T_p\mathcal{Q} \quad \text{and} \quad M^\flat : T_p\mathcal{Q} \longrightarrow T_p^*\mathcal{Q}. \tag{14}$$

Similarly, we have the equivalent Riemannian problem

$$\nabla_U U = G^\sharp\left[u\right], \quad \text{with} \quad u = \sum_{i=1}^{m} u_i \hat{\theta}^{(i)} \quad \text{with} \quad \hat{\theta}^{(i)} = G^\flat\left[\hat{e}_{(i)}\right]. \tag{15}$$

Note that u is the same as in equation (13) but expressed in terms of a different co-frame, the one corresponding to the Jacobi metric G.

If $\text{span}(\mathcal{D}) = T_p\mathcal{Q}$ then we say the system is *fully-actuated*. Otherwise, we say it is *under-actuated*. Most of the relevant situations in control theory involve under-actuated systems, i.e. those for which $\text{span}(\mathcal{D}) \subset T_p\mathcal{Q}$ [5]. Furthermore, if one provides an input force depending solely on time, controlling the evolution of the system in the configuration space $\mathcal{Q}$, then u is called and *open loop control*[1]. However, the central interest in control theory is in those systems which can regulate themselves against unknown perturbations, that is, those whose control input is responsive to spontaneous variations of the system configuration. This is referred as *closed loop control*. In such case, the control input force depends on the state of the system at any given time, that is, u must be a section of the co-tangent bundle and the control objective is that the integral curves of (13) – or, equivalently, those satisfying (15) – remain "close" to a reference path $\gamma^* \subset \mathcal{Q}$ with some desired equilibrium properties.

To clearly state our Riemannian control problem, we propose a slight modification of Lewis' work [7, 8]. Moreover, without any loss of generality and to keep our argument sufficiently simple, we will restrict ourselves to the case where no external nor gyroscopic forces are present (cf. [9]). Thus, let us define an open loop control system as the triad

$$\Sigma_{\text{ol}} \equiv \{\mathcal{Q}, G_{\text{ol}}, \mathcal{W}\}, \tag{16}$$

where $\mathcal{Q}$ and G_{ol} are the configuration space and the Jacobi metric, respectively; and $\mathcal{W} \subset T^*\mathcal{Q}$ is the control sub-bundle defined at each point as

$$\mathcal{W}_p = \text{span}(F_p), \quad F_p = \left\{\hat{\theta}^{(i)}(p)\right\}_{i=1}^{m} \quad \forall p \in \mathcal{Q}, \tag{17}$$

so that equation (15) above is satisfied for a certain $u \in \mathcal{W}$ given some reference γ^*. The goal is to obtain a pair – the closed loop system –

$$\Sigma_{\text{cl}} = \{\mathcal{Q}, G_{\text{cl}}\}, \tag{18}$$

such that the geodesics of G_{cl} *match* the open loop solutions of (15). Here, we use the term match to indicate that the geodesics of the closed loop metric are not required to coincide at every point with the solutions of the open loop system, but merely that the vector fields share the same singular points together with their equilibrium properties, i.e. that the equilibria of the open loop system are the same as those of the closed loop geodesic vector field and with the same asymtotic behaviour. In such case we say that $u \in \mathcal{W}$ is a stabilizing input for the geometry shaping problem solved by G_{cl} [8]. Using this formulation the stability properties of the closed loop system can be assessed directly by means of the Riemann tensor of G_{cl} through the geodesic deviation equation. Intuitively, unstable regions should correspond to subsets of the configuration space where the eigenvalues of the Riemann tensor take negative values and the geodesics in a congruence are divergent, providing us with a completely geometric and coordinate independent stability criterion.

Notice that the closed loop metric is not required to be a Jacobi metric, that is, it is not necessary to find a function $V_{\text{cl}} = V_{\text{cl}}(p)$ such that G_{cl} be of the form (10). Moreover, it may not even share the same signature with G_{ol} (cf. Remark 4.7 in [9]). Finally, recalling that the geodesics of the closed loop metric correspond to the integral curves of a purely kinetic

[1] In fact, open loop *tracking* control for desired reference motion trajectories could be directly synthesized for a class of under-actuated controllable dynamical systems in absence of uncertainty, e.g. differentially flat systems [6]. In such case, system variables (states and control) can be expressed in terms of a set of flat output variables and a finite number of their time derivatives

Lagrangian vector field with kinetic energy metric $G_{\rm cl}$, and that the open loop metric is a Jacobi metric, G_{cl} must satisfy the partial differential equation

$$G^{\sharp}_{\rm ol}\left[\sum_{i=1}^{m} u_i(U_p)\hat{\theta}^{(i)}(p)\right] = \left(\nabla^{\rm cl} - \nabla^{\rm ol}\right)[U_p, U_p], \tag{19}$$

where $\nabla^{\rm cl}$ and $\nabla^{\rm ol}$ are the Levi-Civita connections of $G_{\rm cl}$ and $G_{\rm ol}$, respectively; we have used the shorthand $\left(\nabla^{\rm cl} - \nabla^{\rm ol}\right)[U_p, U_p]$ to denote the *connection difference tensor*, for every pair (p, U_p) denoting a posible state of the system at point $\tilde{p} \in T\mathcal{Q}$ [cf. equation (8), above]. We have been emphatic on the fact that the stabilizing control input must depend on the state of the system. Equations of the form (19) might be used to define further geometric structures in the space they are defined, as has been done in the case of the geometrization programme of thermodynamics and fluctuation theory developed in [10–12].

In recent years, solutions to equation (19) have been the object of various studies [13, 14]. However, one should be aware that the existence of a solution of (19) satisfying some desired properties might be severely constrained by the topology of $\mathcal{Q}$ [8]. Thus, in general, there is no generic criterion for deciding the solvability of (19). Nevertheless, it has been shown that the case of one degree of under-actuation is fully stabilizable by means of (19) [15]; modulo the difference introduced by using the open loop Jacobi metric in the problem's definition. Interestingly, in such case, equation (19) resembles very closely that of the definition of the general relativistic *elasticity difference tensor* introduced by Karlovini and Samuelsson [16]

$$S = P(\nabla - \tilde{\nabla}) \tag{20}$$

where ∇ and $\tilde{\nabla}$ represent the spacetime and the (*pulled back*) matter space metrics, respectively (cf. [17–20] for a detailed presentation of matter spaces in relativistic elasticity and dissipation) and P denotes the orthogonal projection with respect to a free falling geodesic congruence. Such a tensor has been used to recast the relativistic Euler's equations in terms of the Hadamard Elasticity, a form which is elegant and useful in the study of wave propagation in relativistic elastic media [21]. This provides us with an interesting link between relativistic elasticity and geometric control theory which deserves further exploration.

Let us close this contribution by noting that there might be several stabilizing inputs for a given desiderata. In such case, one might look for the 'most suitable' geometry solving our control objective and use (19) as a constraint for a variational problem. Its particular form should be fixed by some *a priori* known cost functional that is to be extremized by the sought for closed loop metric. Such is the standard *optimal control problem* [22]. To preserve the geometric nature of the whole construction, the cost functional can only depend on scalars formed from the metric and its derivatives, that is

$$\mathcal{A}[G] = \int_{\mathcal{Q}} \mathcal{F}\left(G, G', G'', ...\right) \sqrt{G}\, {\rm d}^n q, \tag{21}$$

where $\sqrt{G}\, {\rm d}^n q$ is the invariant volume element on $\mathcal{Q}$. Thus, our search for geometries solving a control objective has led us to the study of cost functionals akin to the various classes of gravitational theories where the extremum is achieved by geometries required to be compatible with the *observed* free fall motion of certain spacetime observers [23]. Therefore, a complete solution to our problem – if it exists –, should be a metric $G_{\rm cl}$ extremizing (21) such that equation (19) is satisfied for a given open loop control. An exploration of the equations of motion stemming from the class of cost functionals constructed from curvature invariants in control theory will be the subject of further investigations. In this sense, the lessons learned and the results obtained from the variational formulation of gravitational theories might find a novel application in the realm of geometric control theory.

Acknowledgements
CSLM wishes to express his gratitude to the Organizing Committee for their kind hospitality and strong efforts in providing us with such an inspiring venue for this celebration.

References

[1] H. Reichenbach. *The Philosophy of Space and Time*. Dover Books on Physics. Dover Publications, 2012.
[2] P.J. Olver. *Applications of Lie Groups to Differential Equations*. Applications of Lie Groups to Differential Equations. Springer New York, 2000.
[3] M. Pettini. *Geometry and Topology in Hamiltonian Dynamics and Statistical Mechanics*. Interdisciplinary Applied Mathematics. Springer New York, 2007.
[4] Andrew D Lewis. Notes on energy shaping. In *Proceedings of the 43rd IEEE Conference on Decision and Control*, volume 5, pages 4818–4823. Citeseer, 2004.
[5] Dave Auckly, Lev Kapitanski, and Warren White. Control of nonlinear underactuated systems. *arXiv preprint math/9901140*, 1999.
[6] Michel Fliess, Jean Lévine, Philippe Martin, and Pierre Rouchon. Flatness and defect of non-linear systems: introductory theory and examples. *International journal of control*, 61(6):1327–1361, 1995.
[7] Francesco Bullo and Andrew D Lewis. *Geometric control of mechanical systems: modeling, analysis, and design for simple mechanical control systems*, volume 49. Springer Science & Business Media, 2004.
[8] Andrew D. Lewis. Is it worth learning differential geometric methods for modeling and control of mechanical systems? *Robotica*, 25(6):765–777, November 2007.
[9] Bahman Gharesifard, Andrew D Lewis, Abdol-Reza Mansouri, et al. A geometric framework for stabilization by energy shaping: Sufficient conditions for existence of solutions. *Communications in Information & Systems*, 8(4):353–398, 2008.
[10] A Bravetti and CS Lopez-Monsalvo. Para-sasakian geometry in thermodynamic fluctuation theory. *Journal of Physics A: Mathematical and Theoretical*, 48(12):125206, 2015.
[11] Alessandro Bravetti, Cesar S Lopez-Monsalvo, and Francisco Nettel. Conformal gauge transformations in thermodynamics. *Entropy*, 17(9):6150–6168, 2015.
[12] P Fernández de Córdoba and JM Isidro. Generalised complex geometry in thermodynamical fluctuation theory. *Entropy*, 17(8):5888–5902, 2015.
[13] N Crasta, Romeo Ortega, and Harish K Pillai. On the matching equations of energy shaping controllers for mechanical systems. *International Journal of Control*, 88(9):1757–1765, 2015.
[14] Wai Man Ng, Dong Eui Chang, and George Labahn. Energy shaping for systems with two degrees of underactuation and more than three degrees of freedom. *SIAM Journal on Control and Optimization*, 51(2):881–905, 2013.
[15] Bahman Gharesifard. Stabilization of systems with one degree of underactuation with energy shaping: a geometric approach. *SIAM Journal on Control and Optimization*, 49(4):1422–1434, 2011.
[16] Max Karlovini and Lars Samuelsson. Elastic stars in general relativity: I. foundations and equilibrium models. *Classical and Quantum Gravity*, 20(16):3613, 2003.
[17] B Carter and H Quintana. Foundations of general relativistic high-pressure elasticity theory. In *Proceedings of the Royal Society of London A: Mathematical, Physical and Engineering Sciences*, volume 331, pages 57–83. The Royal Society, 1972.
[18] César Simón López-Monsalvo. Covariant thermodynamics and relativity. *arXiv preprint arXiv:1107.1005*, 2011.
[19] Cesar S Lopez-Monsalvo and Nils Andersson. Thermal dynamics in general relativity. In *Proceedings of the Royal Society of London A: Mathematical, Physical and Engineering Sciences*, volume 467, pages 738–759. The Royal Society, 2011.
[20] Nils Andersson and Gregory L Comer. Relativistic fluid dynamics: physics for many different scales. *Living Rev. Relativity*, 10(1), 2007.
[21] EGLR Vaz and Irene Brito. Analysing the elasticity difference tensor of general relativity. *General Relativity and Gravitation*, 40(9):1947–1966, 2008.
[22] Suresh P Sethi and Gerald L Thompson. *What is Optimal Control Theory?* Springer, 2000.
[23] David Lovelock. The einstein tensor and its generalizations. *Journal of Mathematical Physics*, 12(3):498–501, 1971.

BTZ black hole assuming running couplings

Ángel Rincón, Benjamin Koch, Ignacio Reyes
Physics Institute, Pontifical Catholic University of Chile, Av. Vicuña Mackenna 4860, Santiago, Chile
E-mail: arrincon@uc.cl

Abstract. In the present work a generalization of the BTZ black hole is studied, for the case of scale dependent couplings. One starts by using the effective action for scale dependence couplings to get a generalization of the Einstein field equations. Self consistent solutions for lapse function, cosmological coupling and Newtons coupling are found. The effect of scale dependent couplings with respect to the classical solution is shown. Moreover, asymptotic behavior as well as thermodynamic properties were investigated. Finally, an alternative way to get the scale dependent Newton coupling, from the so-called "Null Energy Condition" is presented.

1. Introduction

Great effort has been made to try to unify general relativity with quantum mechanics. Two well known candidates for this unification are String theory [1,2] and Loop quantum gravity [3]. Another promising framework for quantum gravity arises from the so called Asymptotic safety scenario [4–9], in which the couplings do not need to be small or tend to zero in the high energy limit. For this scenario to work it necessary to have a finite number of couplings that have a stable UV fixed point. Evidence for the existence of such a non-trivial UV fixed point has been found by the use of non-perturbative RG equations [10]. Independent of the particular RG approach, the outcome of such calculations will give an effective gravitational theory with scale dependence couplings. As simpler toy model for gravity with scale dependent couplings we study three dimensional gravity, which allows for many non-trivial effects despite of the fact that there are no classical propagating degrees of freedom. For example it is known that this theory has a close connection with Chern-Simons theory [11,12]. Another particularly interesting feature of this theory is that there are non-trivial black hole solutions with a negative cosmological constant, found by Bañados, Teitelboim and Zanelli (BTZ) [13,14]. We investigate this black hole in light of the possibility of scale dependent couplings, such as they arise in the asymptotic safety approach. Using the Einstein-Hilbert action, where the couplings are allowed to vary with respect to an arbitrary scale, we derive generalized field equations for the metric and the scale field. Solutions for those equations are obtained by using the schwarzschild ansatz.
The paper is organized as follows: In Section 2 the fundamental ideas, as well as the required framework are discussed. In section 3 the main results are collected. The consequences of our solution are discussed and collected in sections 4 and 5.

2. Main idea

Pioneer works in which quantum gravity is investigated via the effective average action are found at Ref. [15,16]. In this work, the asymptotic safety ideas previously commented are used.

Starting from the Einstein Hilbert action with cosmological constant one take advantage of the Einstein field equation to obtain the equation of motion (hereafter E.O.M.). Thus the effective action, independent of the kind of matter, is

$$\Gamma_k[g_{\mu\nu}, k] = \int d^3x\sqrt{-g}\left[\frac{R - 2\Lambda_k}{G_k}\right], \tag{1}$$

one obtains

$$G_{\mu\nu} + \Lambda_k g_{\mu\nu} = 8\pi G_k T_{\mu\nu}, \tag{2}$$

$$R\frac{\partial}{\partial k}\left(\frac{1}{G_k}\right) = 2\frac{\partial}{\partial k}\left(\frac{\Lambda_k}{G_k}\right), \tag{3}$$

where the effective energy-momentum tensor is given by

$$8\pi G_k T_{\mu\nu} = 8\pi G_k T^m_{\mu\nu} - \Delta t_{\mu\nu}. \tag{4}$$

Note that $T^m_{\mu\nu}$ is the energy-momentum tensor associated with matter and $\Delta t_{\mu\nu}$ reads

$$\Delta t_{\mu\nu} = G_k\Big(g_{\mu\nu}\Box - \nabla_\mu\nabla_\nu\Big)G_k^{-1}. \tag{5}$$

Please, note that k is an arbitrary renormalization scale. If one is interested in purely spherically symmetric settings, one knows that $\mathcal{Q} \equiv \mathcal{Q}\big(k(r)\big) \rightarrow \mathcal{Q}(r)$, where $\mathcal{Q}$ symbolizes unknown functions of the problem e.g. $\{f(r), \Lambda(r), G(r)\}$. Thus, one only needs to solve the E.O.M. for the radial coordinate r [17–19], if one chooses to eliminate one of those functions by a suitable ansatz or a physically motivated condition.

3. Results

One starts by considering a metric in (2+1) dimensional case

$$ds^2 = -f(r)\,dt^2 + g(r)\,dr^2 + r^2 d\phi^2, \tag{6}$$

where $f(r)$ is the lapse function. For vanishing angular momentum ($J_0 = 0$), the so-called "Null Energy Condition" [20] gives a relation between $f(r)$ and $g(r)$ such that $g(r) = 1/f(r)$. This condition is inspired by the Jacobson argument in which one assumes a specific null vector ℓ^μ with $R_{\mu\nu}\ell^\mu\ell^\nu = 0$ [21]. Moreover, according with the E.O.M. (2) one sees that

$$\Delta t_{\mu\nu}\ell^\mu\ell^\nu = 0, \tag{7}$$

which allows to find gravitational scale dependent coupling without actually solve (2)

$$G(r) = \frac{G_0^2}{G_0 + (1 + G_0 M_0)\epsilon r}. \tag{8}$$

Here the constants $\{G_0, M_0, \epsilon\}$ were chosen such that (8) coincides with the classical BTZ solution for $J_0 = 0$. Defining $\delta(r)$ as a contrast to Newtons constant

$$\delta(r) = \frac{G_0}{G(r)} - 1, \tag{9}$$

one can express the generalized solutions for $f(r)$ and $\Lambda(r)$ using (9) and solving the Eq. (2)

$$f(r) = f_0(r) + 2M_0G_0\delta(r)\left[1+\delta(r)\ln\left(\frac{\delta(r)}{1+\delta(r)}\right)\right], \tag{10}$$

$$\Lambda(r) = -\frac{1}{\ell_0^2}\left(\frac{1}{1+\delta(r)}\right)^2\left[1+4\delta(r)+\left(5M_0G_0\frac{\ell_0^2}{r^2}+3\right)\delta(r)^2+6M_0G_0\frac{\ell_0^2}{r^2}\delta(r)^3 + 2M_0G_0\frac{\ell_0^2}{r^2}\Big(1+\delta(r)\Big)\Big(3\delta(r)+1\Big)\delta(r)^2\ln\left(\frac{\delta(r)}{1+\delta(r)}\right)\right]. \tag{11}$$

The set of solutions $\{G(r), f(r), \Lambda(r)\}$ could be either expressed in terms of ϵ parameter (which measures the deviation from the classical solution) or, alternatively, in terms of $\delta(r)$ which measures the same effect but in a more intuitive way. Thus, when $\delta(r) = 0$ the classical case is recovered, and when $\delta(r) \neq 0$ improved solutions (which show the effect of scale dependent couplings) are obtained. In Fig. 1 the lapse function $f(r)$ is shown for different values of ϵ.

3.1. Horizons

In order to have a well-defined black hole surface, a horizon must exist. Horizons are crucial for the understanding of the thermodynamics properties of the black hole. The aparent horizon is found as solution of $f(r_H) = 0$. An analytic expression for the case of r_H can not be obtained. However in certain limits such as $\epsilon \ll 1$ or when $G(r)/G_0$ is small and, simultaneously M_0G_0 is large, analytic expressions are found. Check table (1) to see details. In Fig. (2) one can see the scale dependent effect on the horizon structure of the generalized BTZ solution.

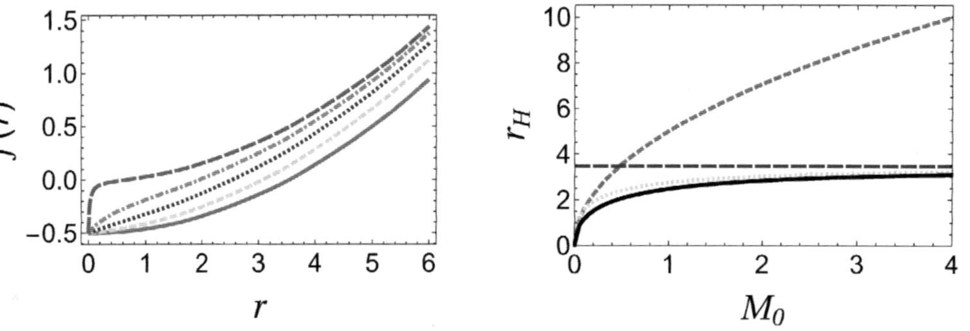

Figure 1. Radial dependence of the lapse function $f(r)$ for $\ell_0 = 5$, $G_0 = 1$ and $M_0 = 0.5$. The different curves correspond to the classical case $\epsilon = 0$ solid red line, $\epsilon = 0.04$ short dashed green line, $\epsilon = 0.15$ dotted blue line, $\epsilon = 0.5$ dot-dashed magenta line, and $\epsilon = 20$ long dashed brown line.

Figure 2. Horizon r_H as a function of M_0 for $\ell_0 = 5$, $G_0 = 1$ and $\epsilon = 0.4$. The curves correspond to the classical case (short dashed red line) for small G/G_0 (dotted green line), for small G/G_0 and large G_0M_0 (long dashed blue line), and the numerical solution (solid black line).

3.2. Thermodynamics

The Bekenstein-Hawking (BH) entropy as well as Hawking temperature are revisited. The usual relation for BH entropy is $S_0 = A/4G$. In the scale dependent case this relation reads [22–26]

$$S = S_0 \cdot \left[1 + \frac{(1+M_0G_0)\epsilon r}{G_0}\right]\Bigg|_{r_H}. \tag{12}$$

The Hawking temperature is obtained by the usual relation $T_H = f'(r_H)/4\pi$. This condition remains even for running couplings. However, since the horizons experience a shift due to $\epsilon \neq 0$, also the Hawking temperature does,

$$T_H = \frac{1}{2\pi r} \frac{G_0^2 M_0}{G_0 + (1 + G_0 M_0)\epsilon r}\bigg|_{r_H} . \tag{13}$$

The running coupling effect on S and T_H is shown in Fig. (3) and Fig. (4). As mentioned before, those numerical results can be approximate analytically in certain regimens which are summarized in the Table 1. Finally, further details about this solution as well as alternative discussion could be found in Ref. [27].

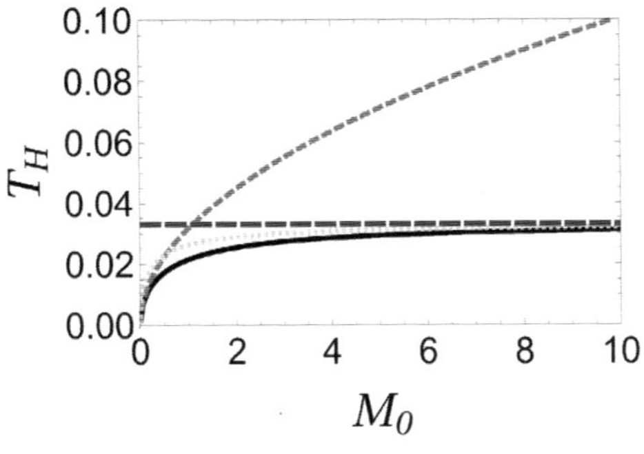

Figure 3. Temperature T_H as a function of M_0 for $\ell_0 = 5$, $G_0 = 1$, and $\epsilon = 0.4$. The curves correspond to the classical case (short dashed red line), for small G/G_0 (dotted green line), for small G/G_0 and large $G_0 M_0$ (long dashed blue line), and the numerical solution (solid black line).

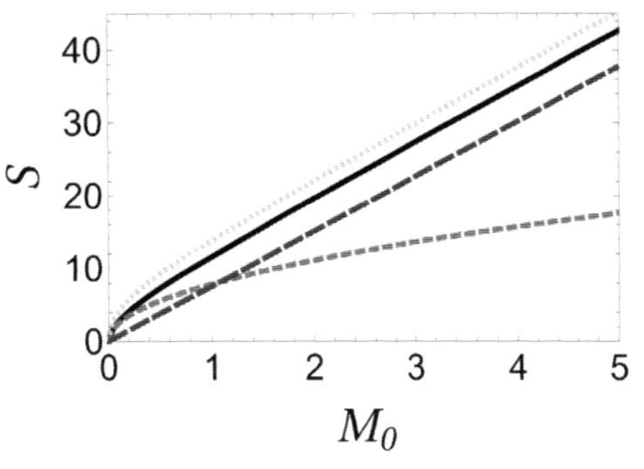

Figure 4. Entropy S as a function of M_0 for $\ell_0 = 5$, $G_0 = 1$, and $\epsilon = 0.4$. The curves correspond to the classical case (short dashed red line), for small G/G_0 (dotted green line), for small G/G_0 and large $G_0 M_0$ (long dashed blue line), and the numerical solution (solid black line).

Table 1. Black hole relevant quantities at different approximations.

Approximation	Entropy S	Temperature T_H	Horizon r_H
$\epsilon \ll 1$	$\frac{\pi}{2}\sqrt{\frac{M_0}{G_0}}\ell_0$	$\frac{\sqrt{M_0 G_0}}{2\pi\ell_0}$	$\sqrt{G_0 M_0}\ell_0$
$G(r)/G_0 \ll 1$	$\pi\left[\frac{\ell_0^4 M_0^2(1+M_0G_0)\epsilon}{18G_0^2}\right]^{\frac{1}{3}}$	$\frac{1}{4\pi}\left[18\frac{M_0G_0^2}{\ell_0^4(1+G_0M_0)\epsilon}\right]^{\frac{1}{3}}$	$\left[\frac{2}{3}\frac{M_0G_0^2\ell_0^2}{(1+M_0G_0)\epsilon}\right]^{\frac{1}{3}}$
$G(r)/G_0 \ll 1$ $M_0G_0 \gg 1$	$\pi M_0\left[\frac{\ell_0^4\epsilon}{18G_0}\right]^{\frac{1}{3}}$	$\frac{1}{4\pi}\left[18\frac{G_0}{\ell_0^4\epsilon}\right]^{\frac{1}{3}}$	$\left[\frac{2}{3}\frac{G_0\ell_0^2}{\epsilon}\right]^{\frac{1}{3}}$

4. Discussion

Due the logarithmic contribution in Eq. (10) it is impossible to find an exact expression for the horizon. This logarithmic term is induced by the scale dependent effect associated with the additional object $\Delta t_{\mu\nu}$. However, it is possible to show that the scale dependent framework does not introduce a new horizon and, therefore, only one horizon appears. The BH entropy grows monotomously with M_0. In addition, the temperature decreases when M_0 goes to zero and tends to a finite value when M_0 goes to infinity.

5. Conclusion

To summarize, in this article the BTZ black hole is investigated in the light of scale dependence. Analytical expressions for the lapse function, the cosmological coupling and, the Newton coupling are determined by solving the Einstein field equations. In figures 1, 2, 3 and 4, the scale dependent effect on the classical BTZ black hole is shown. Due the scale dependence, there appears an effective energy momentum tensor $\Delta t_{\mu\nu}$ associated with the running Newton coupling. According to the "null energy condition" (which reads $R_{\mu\nu}\ell^\mu\ell^\nu = 0$) the usual ansatz $g_{tt} = -1/g_{rr} \equiv f(r)$ is derived i.e. the so called "Schwarzschild relation" is preserved. Special attention is dedicated to the interpretation of the integration constants which is given in terms of the classical parameters $\{G_0, \ell_0, M_0\}$ and one additional constant ϵ, that parametrizes the strength of scale dependence. Horizons, as well as black hole thermodynamics, are investigated. It is found that the large r asymptotic is AdS_3 and that the $r \to 0$ asymptotic has a singular behaviour, not present in the usual BTZ solution. Furthermore, it is found that for fixed values of $\{\epsilon, G_0, \ell_0\}$ the horizon radius saturates for $M_0 \to \infty$ to a finite value. Finally, an important modification of the classical result is found: when $G(r)$ deviates strongly from G_0 a transition from the standard relation $S_0 = A/4G$ to $S = S_0 \times (1 + \alpha r)$ is found. This apparent deviation from the holographic principle is an unusual feature of this new black hole solution using scale dependent couplings.

Acknowledgments

The author A.R. was supported by the CONICYT-PCHA/Doctorado Nacional/2015-21151658. The work of B.K. was supported by the Fondecyt 1161150 and the work of I.R. was partially funded by CONICYT PCHA/Doctorado Nacional # 2015149744.

References

[1] Polchinski J 2007 *String theory. Vol. 1: An introduction to the bosonic string* (Cambridge University Press)
[2] Polchinski J 2007 *String theory. Vol. 2: Superstring theory and beyond* (Cambridge University Press)
[3] Rovelli C 2007 *Quantum gravity* Cambridge Monographs on Mathematical Physics (Cambridge University Press)
[4] Oriti D 2009 *Approaches to quantum gravity: toward a new understanding of space, time and matter* Osborne (Cambridge University Press)
[5] Reuter M and Saueressig F 2012 *New J. Phys.* **14** 055022 (*Preprint* `1202.2274`)
[6] Weinberg S 1979 *Cambridge University Press.* **chapter 16** 790–831
[7] Wetterich C 1993 *Phys. Lett.* **B301** 90–94
[8] Dou D and Percacci R 1998 *Class. Quant. Grav.* **15** 3449–3468 (*Preprint* `hep-th/9707239`)
[9] Reuter M and Saueressig F 2002 *Phys. Rev.* **D65** 065016 (*Preprint* `hep-th/0110054`)
[10] Niedermaier M and Reuter M 2006 *Living Reviews in Relativity* **9** URL `http://www.livingreviews.org/lrr-2006-5`
[11] Witten E 2007 (*Preprint* `0706.3359`)
[12] Witten E 1988 *Nucl. Phys.* **B311** 46
[13] Banados M, Teitelboim C and Zanelli J 1992 *Phys. Rev. Lett.* **69** 1849–1851 (*Preprint* `hep-th/9204099`)
[14] Banados M, Henneaux M, Teitelboim C and Zanelli J 1993 *Phys. Rev.* **D48** 1506–1525 [Erratum: Phys. Rev.D88,069902(2013)] (*Preprint* `gr-qc/9302012`)
[15] Reuter M 1998 *Phys. Rev.* **D57** 971–985 (*Preprint* `hep-th/9605030`)
[16] Reuter M and Weyer H 2004 *Phys. Rev.* **D69** 104022 (*Preprint* `hep-th/0311196`)

[17] Koch B and Ramirez I 2011 *Class. Quant. Grav.* **28** 055008 (*Preprint* 1010.2799)
[18] Koch B and Rioseco P 2016 *Class. Quant. Grav.* **33** 035002 (*Preprint* 1501.00904)
[19] Koch B, Rioseco P and Contreras C 2015 *Phys. Rev.* **D91** 025009 (*Preprint* 1409.4443)
[20] Rubakov V A 2014 *Phys. Usp.* **57** 128–142 (*Preprint* 1401.4024)
[21] Jacobson T 2007 *Class. Quant. Grav.* **24** 5717–5719 (*Preprint* 0707.3222)
[22] Jacobson T, Kang G and Myers R C 1994 *Phys. Rev.* **D49** 6587–6598 (*Preprint* gr-qc/9312023)
[23] Iyer V and Wald R M 1995 *Phys. Rev.* **D52** 4430–4439 (*Preprint* gr-qc/9503052)
[24] Visser M 1993 *Phys. Rev.* **D48** 5697–5705 (*Preprint* hep-th/9307194)
[25] Creighton J D E and Mann R B 1995 *Phys. Rev.* **D52** 4569–4587 (*Preprint* gr-qc/9505007)
[26] Kang G 1996 *Phys. Rev.* **D54** 7483–7489 (*Preprint* gr-qc/9606020)
[27] Koch B, Reyes I A and Rincón A 2016 *Class. Quant. Grav.* **33** 225010 (*Preprint* 1606.04123)

IOP Conf. Series: Journal of Physics: Conf. Series **831** (2017) 012008 doi:10.1088/1742-6596/831/1/012008

Monoid spaces and linearized gravity

E Contreras[12], C Di Bartolo[2] and L Leal[23]

[1]Departamento de Física, Universidad de los Andes, Apartado Aéreo *4976*, Bogotá, Distrito Capital, Colombia
[2]Departamento de Física, Universidad Simón Bolívar, Apartado Postal 89000, Caracas 1080-A, Venezuela.
[3]Centro de Física Teórica y Computacional, Facultad de Ciencias, Universidad Central de Venezuela, AP 47270, Caracas 1041-A, Venezuela.

E-mail: ej.contreras@uniandes.edu.co, dibarto@usb.ve, lleal@fisica.ciens.ucv

Abstract. We propose an alternative representation for linear quantum gravity. It is based on the use of a structure that bears some resemblance to the Abelian loop representation used in electromagnetism but with the difference that the space of extended object on which waves functions take values has a structure of commutative monoide instead of Abelian group. The generator of duality of the theory is realized in this representation and a geometrical interpretation is discussed.

The introduction of loop representation has opened a new avenue for quantization of gauge theories such as electromagnetism [1, 2], Yang-Mills [3] and General Relativity [4, 5, 6, 7, 8, 9, 10, 11, 12, 13]. This representation allows an elegant way of solve the constraints of the theories as well as obtain certain geometrical information about the space of extended objects on which the wave functions take values. For example, it is well known that the Abelian loop representation in electromagnetism allows to solve the Gauss constraint immediately [1, 2]. Also, because the metric independence of the generator of duality, their realization in this representation leads to a knot invariant, more precisely, the Gauss linking number between Abelian loops [2]. Within the same spirit, the loop representation has been used to quantize linearized gravity [14, 15]. For example in reference [15] it was obtained a representation in terms of skein of loops. Although it could be found a realization for the canonical algebra and constraints of the theory were solved in the reduced phase space, a geometrical interpretation of skein representation keeps unclear.

In this paper we develop an alternative representation for linearized quantum gravity. As we shall see, the wave functions in this representation does not depend on a list of three tangled Abelian loops as in reference [15] but in only one extended object. However, the price which has to be paid is that the space of this extended objects does not form an Abelian group but a commutative monoid space.

We start by considering a set of parametrized curves γ on R^3 given by $\vec{z}_\gamma = \vec{z}(l)$, with l the arc length parameter, and define

$$I^{ab}[\vec{x},\gamma] := \int_\gamma dl u^a_{T_\gamma} u^b_{T_\gamma} \delta^3(\vec{z}_\gamma - \vec{x}), \tag{1}$$

with $\hat{u}_{T_\gamma} := d\vec{z}_\gamma/dl$ the tangent vector to γ and $a, b = 1, 2, 3$. It is easy to check that (1) is symmetric and independent of the curve orientation. It is worth mentioning that for an arbitrary parameter τ we have

$$I^{ab}[\vec{x}, \Gamma] = \int_{\tau_1}^{\tau_2} d\tau \frac{dz^a(\tau)}{d\tau} \frac{dz^b(\tau)}{d\tau} \left| \frac{d\vec{z}(\tau)}{d\tau} \right|^{-1} \delta^3(\vec{x} - \vec{z}(\tau)), \tag{2}$$

with $\tau_1 < \tau_2$ independent of the orientation of γ.

Let $\mathcal{M}$ be the space which elements Γ are the union of disjoint curves, i.e, $\Gamma = \gamma_1 \cup .... \cup \gamma_n$ with arbitrary n. From definition (1), it follows that

$$I^{ab}[\vec{x}, \Gamma] = I^{ab}[\vec{x}, \gamma_1 \cup ... \cup \gamma_n] = \sum_{i=1}^{n} I^{ab}[\vec{x}, \gamma_i]. \tag{3}$$

Now, we shall say that two curves Γ and Γ' belong to the same equivalence class if $I^{ab}[\vec{x}, \Gamma] = I^{ab}[\vec{x}, \Gamma']$ and hereafter, they will represent the same element or *path*. We will denote the space formed by such a paths by $\mathfrak{M}$.

For elements in $\mathfrak{M}$ we define the product $\circ$ among two paths Γ_1 and Γ_2 as follows

$$I^{ab}[\vec{x}, \Gamma_1 \circ \Gamma_2] := I^{ab}[\vec{x}, \Gamma_1 \cup \Gamma_2] = I^{ab}[\vec{x}, \Gamma_1] + I^{ab}[\vec{x}, \Gamma_2]. \tag{4}$$

As can be seen from the above definition, this product is commutative. In the other hand, for an element $\Gamma_e \in \mathfrak{M}$ such as $I^{ab}[\vec{x}, \Gamma_e] = 0$ (the null path) occurs that

$$I^{ab}[\vec{x}, \Gamma \circ \Gamma_e] = I^{ab}[\vec{x}, \Gamma]. \tag{5}$$

Hence Γ_e is the neutral element for the multiplication defined in (4). In summary, the elements of $\mathfrak{M}$ along with (4) form a commutative monoid, i.e, a semi-group with an identity.

Let us now consider square integrable functions (at least formally) $\Psi : \mathfrak{M} \to C$. We define the path derivative operator $\delta_{cd}(\vec{y}, \hat{v})$ as

$$v^c v^d \delta_{cd}(\vec{y}, \hat{v}) \Psi[\Gamma] := \lim_{L \to 0} \frac{1}{L} \left(\Psi[\Gamma] \circ \Gamma[\vec{y}, \hat{v}, L] - \Psi[\Gamma] \right). \tag{6}$$

As can be seen, this derivative compute the change of $\Psi[\Gamma]$ when a path $\Gamma[\vec{y}, \hat{v}, L]$ of infinitesimal length L starting at $\vec{y}$ in the $\hat{v}$ direction is appended to the path Γ. Now with the above definition, is straightforward to prove that the path derivative for $I^{ab}[\vec{x}, \Gamma]$ is given by

$$\delta_{cd}(\vec{y}, \hat{v}) I^{ab}[\vec{x}, \Gamma] = \frac{1}{2}(\delta^a_c \delta^b_d + \delta^a_d \delta^b_c) \delta^3(\vec{x} - \vec{y}). \tag{7}$$

With this tools at hand we are ready to obtain a monoidal representation for linearized gravity. Before doing that, let us recall the most relevant aspects of linearized gravity at Hamiltonian level. In the first place the canonical variables are the spatial components of the metric perturbation field h_{ab} and this conjugated momentum p^{ab} with Poisson algebra

$$\{h_{ab}(\vec{x}), p^{cd}(\vec{y})\} = \frac{1}{2}(\delta^c_a \delta^d_b + \delta^c_b \delta^d_a) \delta^3(\vec{x} - \vec{y}). \tag{8}$$

Also, there are two first class constraints given by

$$\partial_a p^{ab} \approx 0$$
$$\partial^a \partial^b h_{ab} - \nabla^2 h \approx 0, \tag{9}$$

where h is the trace of h_{ab}. As has been discussed in reference [15], the constraints (9) lead to conclude that observables of the theory must be the transverse and traceless part of the canonical pair, in the same way that Gauss constraint in electromagnetism leads to the observable in that case is the transverse part of the electric field.

In order to obtain the quantum theory for linearized gravity, we shall proceed to implement the Dirac quantization program as described below. The canonical variables are promoted to operators on certain Hilbert space

$$\begin{aligned} h_{ab} &\rightarrow \hat{h}_{ab} \\ p^{ab} &\rightarrow \hat{p}^{ab}, \end{aligned} \tag{10}$$

with commutator given by

$$[\hat{h}_{ab}(\vec{x}), \hat{p}^{cd}(\vec{y})] \rightarrow i\{h_{ab}(\vec{x}), p^{cd}(\vec{y})\} = \frac{i}{2}(\delta_a^c \delta_b^d + \delta_b^c \delta_a^d)\delta^3(\vec{x} - \vec{y}). \tag{11}$$

The physical states are those annihilated by the first class constraints

$$\partial_a \hat{p}^{ab} |\Psi >= 0$$
$$(\partial^a \partial^b \hat{h}_{ab} - \nabla^2 \hat{h})|\Psi >= 0, \tag{12}$$

and their time evolution is given by the Schrödinger equation.

Now, to obtain the quantum theory for linearized gravity we must represent the operators (10) acting on $\Psi[\Gamma]$ in such a way that they fulfill the canonical algebra (11) and solve the first class constraints (12). With the purpose of carrying out the quantization previously described, let us consider the follow realization

$$\begin{aligned} \hat{h}_{ab}\Psi[\Gamma] &\rightarrow i\delta_{ab}(\vec{x}, \hat{v})\Psi[\Gamma] \\ \hat{p}^{ab}\Psi[\Gamma] &\rightarrow I^{ab}[\vec{x}, \Gamma]\Psi[\Gamma]. \end{aligned} \tag{13}$$

After a straightforward calculation involving derivative (7) it can be shown that this prescription fulfill the canonical algebra (11). Regrettably, unlike electromagnetism case in the Abelian loop representation, the first class constraints can not be realized in a direct way. However, we can consider the quantization of the theory in the reduce phase space, i.e, taking the transverse and traceless part of the canonical variables as described in reference ([15]). Let $\hat{h}_{ab}^{TT}$ and $\hat{p}^{abTT}$ be the transverse and traceless part of the dynamical field and their conjugate moment respectively and $P_{ab/cd} = P_{ac}P_{db} - \frac{1}{2}P_{ab}P_{cd}$ be the transverse and traceless part projector (P_{ab} is the usual transverse projector of electromagnetism [15, 16, 17]). The transverse and traceless parts of our original operators are given by

$$\begin{aligned} \hat{h}_{ab}^{(TT)} &= P_{ab/cd}\hat{h}_{cd} \\ \hat{p}_{ab}^{(TT)} &= P_{ab/cd}\hat{p}^{cd}. \end{aligned} \tag{14}$$

IOP Conf. Series: Journal of Physics: Conf. Series **831** (2017) 012008 doi:10.1088/1742-6596/831/1/012008

It worth mentioning that considering (14) as our basics quantum operators, constraints (9) are "strong" equalities.

At this stage we can ask if our results (13) admits an easy geometrical interpretation as in electromagnetism case. In fact, Abelian loops are interpreted as quantum Faraday's lines which are closed in absent of sources and this solve automatically the Gauss constraint. In the other hand, the quantum operator associated to the electric field is interpreted as a measure of the density of electric flux. However, in the monoidal representations that we found such an interpretation is unclear. In the first place, we have not found a solution for the constraints in terms of conditions on the paths space but we have restricted to reduced phase space. Secondly, the observable associated with a kind of density of flux of Faraday's lines is p^{abTT} which have non-local terms that make difficult to extract some geometrical information.

In spite of the former observations, let us consider the monoid representation for the generator of duality [15, 17]

$$\hat{G} = \int d^3x(-\frac{1}{4}\hat{h}^{(TT)}_{ab}\varepsilon^{acd}\partial_c\hat{h}^{(TT)b}_{d} + \hat{p}^{(TT)}_{ab}\nabla^{-2}\varepsilon^{acd}\partial_c\hat{p}^{(TT)b}_{d}). \tag{15}$$

As can be seen, this generator is similar to obtained in the electromagnetic case (see for example [2]) in the sense that it depends only on the observables of the theory combined in a form that "looks like" Chern-Simons terms. But these similarities are only formals because terms in (15) are metric dependents and for this reason we are not expecting that their realization leads to knot invariants. However, some interesting information can be obtained as we shall see in what follow. Performing the representation (13) and the projector (14) in (15) we obtain

$$\begin{aligned}\hat{G} &= \int d^3x\delta_{ab}(\vec{x},\hat{v})\varepsilon^{acd}\partial_c\delta_d{}^b(\vec{x},\hat{v}) + \int d^3x\partial^b\delta_{ab}(\vec{x},\hat{v})\varepsilon^{acd}\partial_c\nabla^{-2}\partial^e\delta_{de}(\vec{x},\hat{v}) \\ &+ \int d^3xI^{ab}[\vec{x},\Gamma]\varepsilon_{acd}\nabla^{-2}\partial^cI^d{}_b[\vec{x},\Gamma] + \int d^3x\partial_bI^{ab}[\vec{x},\Gamma]\varepsilon_{acd}\partial^c\nabla^{-4}\partial_eI^{de}[\vec{x},\Gamma]. \end{aligned}\tag{16}$$

To obtain an explicit form of the generator in terms of the paths Γ we must to use definitions (1) and (6). However in the same way as electromagnetic case the "electric part" of the generator have an interesting interpretation in this representation. In our case it corresponds to the last two terms. For this reason let us consider only the "electric terms". Also, in order to simplify some calculus and to facilitate the geometrical interpretation, we shall assume that each term has support on different closed paths. Taking into account the former considerations, the third term, that we call $\hat{G}_3$, can be expressed as

$$\hat{G}_3 = -\frac{1}{4\pi}\oint_{\Gamma_1} dl_{\Gamma_2}\oint_{\Gamma_2} dl_{\Gamma_2}(\hat{u}_{T_{\Gamma_1}}\cdot\hat{u}_{T_{\Gamma_2}})\int_{\gamma}^{\vec{z}_{\Gamma_1}} dl_\gamma(\hat{u}_{T_{\Gamma_1}}\times\hat{u}_{T_{\Gamma_2}})\cdot\hat{u}_{T_\gamma}\delta^3(\vec{z}_{\Gamma_2}-\vec{w}_\gamma), \tag{17}$$

where the last integral is performed on an auxiliar open straight curve γ from $-\infty$ to $\vec{y}$. Please, note that this term comes from replace

$$-\frac{1}{4\pi}\frac{(\vec{x}-\vec{y})^a}{|\vec{x}-\vec{y}|^3} = \int_{\gamma}^{\vec{y}} dz^a\delta^3(\vec{x}-\vec{y}) + \varepsilon^{abc}\partial_bf_c(|\vec{x}-\vec{y}|), \tag{18}$$

where $f_c(|\vec{x}-\vec{y}|)$ is an arbitrary function and as in electromagnetism case [2], the curl does not contribute to the generator. It can be seen that (17) is a kind of Gauss linking number. The only difference is that it has the term $\hat{u}_{T_{\Gamma_1}} \cdot \hat{u}_{T_{\Gamma_2}}$ which is a measure of the angle between tangent vector of the closed paths. Note that despite of that, $\hat{G}_3$ will have non-vanish contribution when closed path Γ_1 intersect the surface closed by Γ_2. However, instead of obtain ± 1 every time this occurs as in electromagnetism [2], our result is angle dependent, or in other words, is metric dependent as expected. Another way to interpret $\hat{G}_3$ is the following. Note that it will contribute as long as tangent vectors $\hat{u}_{T_{\Gamma_1}}, \hat{u}_{T_{\Gamma_2}}, \hat{u}_{T_\gamma}$ form a trihedron but also the dihedral angle between planes intersecting in the line generated by $\hat{u}_{T_\gamma}$ must be necessarily different from 0 and $\pi/2$.

In the other hand the fourth term, that we call $\hat{G}_4$, is given by

$$\hat{G}_4 = -\left(\frac{1}{4\pi}\right)^2 \oint_{\Gamma_1} dl_{\Gamma_1} \kappa_{\Gamma_1} \oint_{\Gamma_2} dl_{\Gamma_2} \int_{\gamma_1}^{\vec{z}_{\Gamma_1}} dl_{\gamma_1} (\hat{u}_{N_{\Gamma_1}} \times \hat{u}_{T_{\Gamma_2}}) \cdot \hat{u}_{T_{\gamma_1}} \int_{\gamma_2}^{\vec{w_{\gamma_1}}} dl_{\gamma_2} \hat{u}_{T_{\gamma_2}} \cdot \hat{u}_{T_{\Gamma_2}} \delta^3(\vec{w}_{\gamma_2} - \vec{z}_{\Gamma_2}), \quad (19)$$

where $\hat{u}_{N_{\Gamma_1}}$ es the normal vector to Γ_1 and κ_{Γ_1} is their curvature. The interpretation for this term is less direct that the former. However if we consider the auxiliar curves γ_1 and γ_2 be parallel, we can said that this term will contribute a long as ($\hat{u}_{N_{\Gamma_1}}, \hat{u}_{T_{\Gamma_2}}$ and $\hat{u}_{T_{\gamma_1}}$) form a thriedron and also the dihedral angle between planes intersecting in the line generated by $\hat{u}_{N_{\Gamma_1}}$ must be different from 0 and $\pi/2$.

It seems possible to extend the above formulation to treat with another models. For example, linearized massive gravity could also be formulated in the monoidal representation following the scheme of quantization presented here. We want to make it clear that the only advantages of the monoidal representation presented here in comparison with the well known loop representation in the context of linearized gravity, is that it allows to represent symmetric objects (namely the canonical pair (h_{ab}, p^{ab})) with just one kind of extend objects instead of three. We found that this feature allows us to carry out a kind of "geometric interpretation" of interesting objects like the generator of duality of the theory. If the monoidal representation allows to obtain some hints that can be further applied or generalized to full gravity or if can provide more insights respect to the standard loop formulation is under investigation.

References

[1] Di Bartolo C, Nori F, Gambini R and Trias A 1982 *Lett. Nuo. Cim.* **38** 15
[2] Contreras E, Martínez Y and Leal L 2010 *Phys. Rev.* D **82** 105013
[3] Gambini R and Trias A 1986 *Nucl. Phys.* B **278** 436
[4] Rovelli C and Smolin L 1990 *Nucl. Phys.* B **331** 80
[5] Rovelli C and Smolin L 1995 *Phys. Rev.* D **52** 5743
[6] Barbero J 1995 *Phys. Rev.* D **51** 5507
[7] Holst S 1996 *Phys. Rev.* D **53** 5966
[8] Gaul M and Rovelli C 2000 *Lect. Notes Phys.* **541** 277
[9] Ashtekar A 1986 *Phys. Rev. Lett.* **57** 2244
[10] Ashtekar A, Rovelli C and Smolin L 1991 *Phys. Rev.* D **44** 1740
[11] Ashtekar A and Lewandowski J 2004 *Class. Quantum Grav.* **21** R53
[12] Ashtekar A 2005 *New J.Phys.* **7** 198
[13] Ashtekar A 1987 *Phys. Rev.* D **36** 1587
[14] Varadarajan M 2002 *Phys.Rev.* D **66** 024017
[15] Contreras E and Leal L 2014 *Int. J. Mod. Phys.* D **23** 1450047
[16] Misner C, Thorne K and Wheeler J 1973 *Gravitation* (San Francisco: Freeman) pp 946-949
[17] Barnich G and Troessaert G 2009 *J.Math.Phys.* **50** 042301

IOP Conf. Series: Journal of Physics: Conf. Series **831** (2017) 012009 doi:10.1088/1742-6596/831/1/012009

Spherical steady-state accretion of a relativistic collisionless gas into a Schwarzschild black hole

Paola Rioseco and Olivier Sarbach

Instituto de Física y Matemáticas, Universidad Michoacana de San Nicolás de Hidalgo, Edificio C-3, Ciudad Universitaria, 58040 Morelia, Michoacán, México.

E-mail: `prioseco@ifm.umich.mx`, `sarbach@ifm.umich.mx`

Abstract. In previous work, we derived the most general solution of the collisionless Boltzmann equation describing the accretion of a kinetic gas into a Schwarzschild black hole background, and we gave explicit expressions for the corresponding observables (the current density and stress energy-momentum tensor) in terms of certain integrals over the distribution function. In this article, we numerically compute these integrals for the particular case of the steady-state, spherical symmetric accretion flows which, at infinity, are described by an equilibrium distribution function of given temperature. We analyze in detail the behavior of the observables as a function of the temperature and the radial coordinate, comparing our results with the perfect fluid model of Bondi-Michel accretion.

1. Introduction

During the last few years there has been an increasing interest in the relativistic kinetic theory of gases, originating from the need to consider kinetic gases in extreme situations, either at very high temperatures or in the presence of strong gravitational fields, where the standard Newtonian theory breaks down. Such situations typically occur in astrophysical or cosmological scenarios, for instance in the vicinity of black holes or in the early Universe. For a recent book on relativistic kinetic theory, we refer the reader to [1]. Other relevant reviews include the book by Stewart [2], the survey article by Ehlers [3], and the survey article by Andréasson [4] on more recent mathematical applications of relativistic kinetic theory. Incorporating the principle of general covariance of Einstein's theory of general relativity, the general relativistic kinetic theory of gases possesses an elegant geometric formulation based on the tangent (or cotangent) bundle associated with the spacetime manifold (M, g), see for example Refs. [5, 3, 6, 7, 8, 9, 10].

In previous work [10], we provided a systematic study for the propagation of a collisionless, relativistic kinetic simple gas on a nonrotating black hole background, neglecting its self-gravity. To this end, we used tools from the theory of integrable Hamiltonian systems and derived the most general solution of the collisionless Boltzmann equation on a Schwarzschild background describing accretion. Further, we derived explicit expressions for the observables (namely, the current density and the stress energy-momentum tensor) and specialized them to the particular case of a spherical, steady-state flow. Assuming that asymptotically, the gas is isotropic and in thermodynamic equilibrium at some temperature $T > 0$, we computed the accretion and compression rates, the energy density and the radial and tangential pressures at the horizon in the limit in which the thermal energy $k_B T \ll mc^2$ is much smaller than the rest energy mc^2 of

IOP Conf. Series: Journal of Physics: Conf. Series **831** (2017) 012009 doi:10.1088/1742-6596/831/1/012009

the gas particles.[1] In this limit, we found that the tangential pressure at the horizon is about ten times larger than the radial one. This provides a partial explanation for the known fact that the accretion rate for a collisionless gas is low compared to the Bondi-Michel accretion model in the hydrodynamic, isotropic perfect fluid case [11, 12, 13, 14]. Furthermore, in [10] we showed an asymptotic stability result, proving that rather general initial gas configurations (which are not necessarily stationary nor spherically symmetric) relax in time to a spherical steady-state solution.

Our intention here is to analyze the behavior and properties of the observables corresponding to the spherical steady-state models in more detail. Whereas in our previous work we only analyzed the low temperature limit and the behavior of the observables at the horizon and in the asymptotic region, here we extend the analysis to arbitrary temperatures and values of the radial coordinate. This is achieved by computing numerically the integrals over the momenta derived in [10] and plotting the results. By interpreting these plots we obtain a better understanding for the physical properties of a collisionless kinetic gas which is accreted by a black hole and the manner it differs from the behavior of a perfect fluid in the same situation. In the last part of the article, we provide an explanation for the reason why the kinetic gas ceases to behave as an isotropic perfect fluid in the vicinity of the black hole, even though in our model it fulfills the same conditions as a perfect fluid at infinity.

2. The spherical steady-state model

Our (highly idealized) model of black hole accretion is based on the assumptions that the accretion flow is steady-state, spherically symmetric, and that the self-gravity and collisions of the kinetic gas being accreted can be neglected. In the cotangent-bundle formulation the gas configuration can be described by a one-particle distribution function $f(x,p)$ which satisfies the collisionless Boltzmann equation[2]

$$\left(p^\mu \frac{\partial}{\partial x^\mu} - \frac{1}{2} p_\alpha p_\beta \frac{\partial g^{\alpha\beta}}{\partial x^\mu}\frac{\partial}{\partial p_\mu}\right) f = 0 \tag{1}$$

on a Schwarzschild black hole background (M,g). The spacetime observables, that is, the current density and stress energy-momentum tensor, are obtained from the distribution function by integration over the momentum space C_x consisting of those p which have the form $p = g_{\mu\nu}(x)p^\nu dx^\mu$ with $p^\nu\partial_\nu$ denoting a future-directed timelike tangent vector at x:

$$J_\mu(x) = \int_{C_x} p_\mu f(x,p)\mathrm{dvol}_x(p), \tag{2}$$

$$T_{\mu\nu}(x) = \int_{C_x} p_\mu p_\nu f(x,p)\mathrm{dvol}_x(p), \tag{3}$$

with $\mathrm{dvol}_x(p) := \sqrt{-\det(g^{\mu\nu}(x))}d^4p$ the natural volume element on C_x.

In [10] we derived the most general solution of Eq. (1) in terms of appropriate symplectic coordinates on the cotangent bundle T^*M associated with the Schwarzschild spacetime (M,g) and provided a detailed discussion of the symmetries of f. This was achieved by exploiting the fact that geodesic motion in a Schwarzschild spacetime constitutes an integrable Hamiltonian system. By imposing spherical symmetry and stationarity, we showed that any steady-state, spherical configuration is described by a distribution function of the form

$$f(x,p) = \mathcal{F}(m,E,L), \tag{4}$$

[1] As usual, k_B denotes Boltzmann's constant and c the speed of light.

[2] Here, (x^μ, p_μ) refer to adapted local coordinates on the cotangent bundle T^*M, in which the canonical momentum is expanded as $p = p_\mu dx^\mu$.

with $\mathcal{F}$ a smooth function of the mass m, energy E, and total angular momentum L of the gas particles, which, in terms of standard Schwarzschild coordinates $(t, r, \vartheta, \varphi)$ are given by

$$m = \sqrt{-p^\mu p_\mu}, \qquad E = -p_t, \qquad L = \sqrt{p_\vartheta^2 + \frac{p_\varphi^2}{\sin^2\vartheta}}. \tag{5}$$

Since these quantities are conserved along geodesics, it is clear that Eq. (4) provides a solution of the collisionless Boltzmann equation. The arguments presented in [10] show that any steady-state, spherical solution must be of this form.

For the following, we consider the situation in which there is a reservoir of identical particles at infinity. We assume that these particles have positive rest mass $m > 0$ and that their momenta are isotropically distributed. Additionally, we assume that (although the gas is assumed to be collisionless) some physical process in the past drove the reservoir to thermodynamic equilibrium at some fixed temperature $T > 0$. As a consequence of (4), this implies that the distribution function is given by

$$f(x, p) = \alpha\delta(\sqrt{-p^\mu p_\mu} - m)e^{-z\varepsilon}, \tag{6}$$

where α is a positive amplitude with untis $1/(\text{volume} \times \text{mass}^4 \times \text{velocity}^4)$, $\varepsilon := E/(mc^2)$ is the ratio between the total energy of the particle and its rest energy, and z is the dimensionless inverse temperature

$$z := \frac{mc^2}{k_B T}. \tag{7}$$

As stated in the introduction, in [10] we explicitly computed the associated observables at the horizon and in the asymptotic region in the low temperature limit $z \to \infty$, and showed that the accretion rate agrees with known results based on Newtonian calculations [15, 13]. In the following, we extend these results to arbitrary values of z and the dimensionless radial coordinate $\xi := 2r/r_H$, with r_H the event horizon radius.

Even though this accretion model is highly idealized and based on a number of assumptions that eventually need to be relaxed, it exhibits rich phenomenology as we show next.

3. Phenomenology of the spherical steady-state model

Explicit expressions for the particle and energy fluxes, the current density, and the stress energy-momentum tensor for a distribution function of the form (6) in terms of integrals over the variable ε are given in Section IV.C of Ref. [10]. In the following, we compute these integrals numerically and analyze their dependency on z and ξ.

The particle accretion rate (number of particles that cross the horizon per unit time) and the energy accretion rate (total energy crossing the horizon per unit time) are given by

$$\dot{\mathcal{N}} = \alpha m^4 c^5 \pi^2 r_H^2 \int_1^\infty \lambda_c(\varepsilon)^2 e^{-z\varepsilon} d\varepsilon, \tag{8}$$

$$\dot{\mathcal{E}} = \alpha m^5 c^7 \pi^2 r_H^2 \int_1^\infty \varepsilon \lambda_c(\varepsilon)^2 e^{-z\varepsilon} d\varepsilon, \tag{9}$$

respectively. In these expressions,

$$\lambda_c(\varepsilon) = \sqrt{\frac{12}{1 - 4a - 8a^2 + 8a\sqrt{a^2 + a}}}\Bigg|_{a=\frac{9}{8}\varepsilon^2 - 1}, \tag{10}$$

IOP Conf. Series: Journal of Physics: Conf. Series **831** (2017) 012009 doi:10.1088/1742-6596/831/1/012009

is the critical value of the (dimensionless) total angular momentum, below which the particles are absorbed by the black hole. For a derivation of Eq. (10) and details, see Appendix A in [10]. For the purpose of this article it is enough to know that $\lambda_c(\varepsilon)$ is a monotonously increasing function such that $\lambda_c(1) = 4$ and $\lambda_c(\varepsilon) \simeq \sqrt{27}\varepsilon$ for $\varepsilon \to \infty$.

For very low temperatures ($z \to \infty$), $\dot{\mathcal{E}} \simeq mc^2\dot{\mathcal{N}}$, since each particle's energy $E \simeq mc^2$ is dominated by its rest mass energy. However, for finite values of z the energy accretion rate $\dot{\mathcal{E}}$ is larger than $mc^2\dot{\mathcal{N}}$ since it includes the internal energy of the gas. When analyzing the behavior of these accretion rates as a function of temperature, it is convenient to eliminate the unphysical parameter α. This can be done by replacing α with the particle density at infinity, given by [16, 17, 18]

$$n_\infty(z) = 4\pi\alpha(mc)^4\frac{K_2(z)}{z}, \tag{11}$$

where K_2 is the modified Bessel function of the second kind. In Fig. 1 we plot the dimensionless quantities $\dot{\mathcal{N}}/(r_H^2 c n_\infty)$ and $\dot{\mathcal{E}}/(r_H^2 mc^3 n_\infty)$ as a function of the dimensionless inverse temperature z.

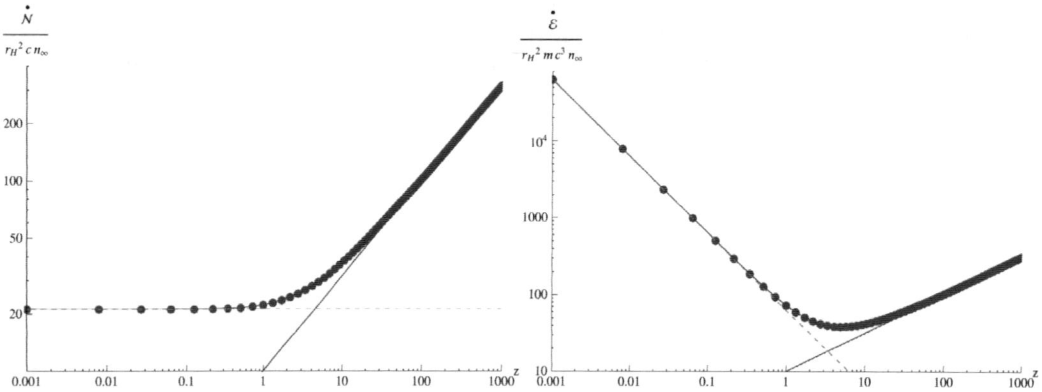

Figure 1. Particle and energy accretion rates as a function of the inverse temperature z. The solid blue lines describe the asymptotic behavior in the low temperature limit $z \to \infty$, see Eqs. (12,13). The red dashed lines describe the behavior in the high temperature limit, see Eqs. (14,15).

For low temperatures one finds [10]

$$\dot{\mathcal{N}} \sim 4r_H^2 c\sqrt{2\pi z}n_\infty \simeq (8.3\times 10^{28})\left(\frac{M_H}{10M_\odot}\right)^2\left(\frac{n_\infty}{1cm^3}\right)\left(\frac{z}{10^9}\right)^{1/2}\frac{1}{s}, \tag{12}$$

$$\dot{\mathcal{E}} \sim 4r_H^2 mc^3\sqrt{2\pi z}n_\infty \simeq (2.2\times 10^{-21})\left(\frac{M_H}{10M_\odot}\right)^2\left(\frac{m}{m_p}\right)\left(\frac{n_\infty}{1cm^3}\right)\left(\frac{z}{10^9}\right)^{1/2}\frac{M_\odot c^2}{yr}, \tag{13}$$

with $M_\odot$ and m_p denoting the solar and proton mass, respectively, and where we have used typical values for the ionized component of the interstellar medium in our Galaxy. This result agrees with Newtonian-based calculations [15, 13], and it differs from the corresponding result in the isotropic perfect fluid case by a factor of the order of z, see for instance [13]. In the opposite limit of extremely high temperatures we find, using a similar method than the one described in Appendix C of [10],

$$\dot{\mathcal{N}} \simeq \frac{27\pi}{4}r_H^2 c n_\infty, \tag{14}$$

$$\dot{\mathcal{E}} \simeq \frac{81\pi}{4} \frac{r_H^2 mc^2 n_\infty}{z}. \tag{15}$$

We see that while the particle accretion rate converges to a constant for small z, the energy accretion rate diverges like $1/z$.

At this point, the question arises why the particle accretion rate $\dot{\mathcal{N}}$ decreases steadily as the temperature increases, while the energy accretion rate $\dot{\mathcal{E}}$ starts decreasing until it reaches a minimum, located about $z \sim 1$, and then increases again. This difference can be understood as follows. For small temperatures most of the particles in the reservoir have very low internal energy, and thus most of the infalling particles are absorbed by the black hole, leading to a high particle accretion rate (compared to $r_H^2 cn_\infty$). In contrast, at high temperatures, only a small fraction of the infalling particles will be absorbed by the hole, since most of the gas particles have large enough angular momenta $\lambda > \lambda_c(\varepsilon)$. To make this point more precise, we consider the critical impact parameter below which the particle is absorbed by the black hole, which is found to be

$$b_c(\varepsilon) = \frac{L_c(E)}{\sqrt{E^2 - m^2}} = \frac{r_H}{2} \frac{\lambda_c(\varepsilon)}{\sqrt{\varepsilon^2 - 1}}. \tag{16}$$

For small temperatures, most particles have energy $\varepsilon \simeq 1$ yielding large values of $b_c(\varepsilon)$, showing that in this case most particles are absorbed by the black hole. In contrast, at high temperatures most particles have energies $\varepsilon \gg 1$ in which case $b_c(\varepsilon) \simeq \sqrt{27} r_H/2$, and only a small fraction of the particles fall into the black hole. However, in this case, the internal energy of each absorbed particle is high, and thus although a small fraction of the particles are absorbed at high temperatures, each of the absorbed particles carries much more energy than in the low temperature case, giving rise to a large energy accretion rate. The behavior of $\dot{\mathcal{E}}$ displayed in the right panel of Fig. 1 is the result of these two competing effects (the smaller fraction of absorbed particles versus higher internal energy for each gas particles as the temperature goes up).

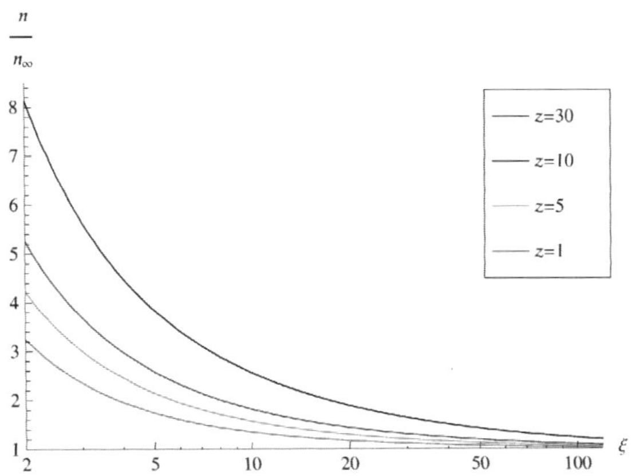

Figure 2. The behavior of the compression rate as a function the dimensionless radial coordinate ξ for different temperatures.

To reinforce this interpretation, we compute the compression rate n/n_∞ and the energy density ρ of the kinetic gas as a function of ξ at different temperatures. In Fig. 2, we show the behavior of the compression rate n/n_∞. Here, the particle density is defined invariantly by

$$n(\xi) = \sqrt{-J^\mu(\xi) J_\mu(\xi)}, \tag{17}$$

with J^μ the current density whose explicit representation can be found in Section IV.C in [10]. As can be observed from the plots in Fig. 2, at fixed temperature the compression rate increases as one moves closer to the black hole, as expected. However, at any fixed position, the compression rate decreases as the temperature increases. This further illustrates the fact that at high temperatures, a smaller fraction of the particles is absorbed by the black hole.

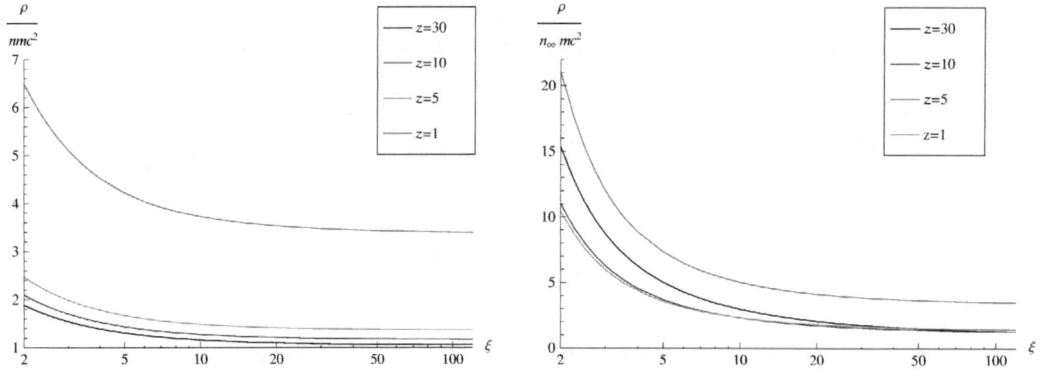

Figure 3. The behavior of the energy density ρ as a function of ξ at different temperatures. Left panel: Ratio between $\rho(\xi)$ and the rest energy density $n(\xi)mc^2$. Right panel: Ratio between $\rho(\xi)$ and the rest energy density $n_\infty mc^2$ at infinity.

Next, in Fig. 3 we show the behavior of the energy density $\rho(\xi)$ of the gas, compared to both its rest energy density at ξ and the rest energy density at infinity. Here, $\rho(\xi)$ is defined invariantly as minus the eigenvalue corresponding to the unique future-directed, unit timelike eigenvector $\mathbf{e}_0$ of $T^\mu{}_\nu(\xi)$, such that $T^\mu{}_\nu e_0^\nu = -\rho e_0^\mu$. ρ is computed numerically from the explicit expressions for $T^\mu{}_\nu(\xi)$ given in Section IV.C of Ref. [10]. We see from the plots in the left panel that as expected, the quantity $\rho(\xi)/(n(\xi)mc^2)$ (being equal to one plus the fraction between the internal energy of the gas and its rest energy) increases as one moves closer to the black hole or as the temperature goes up. In contrast to this, the behavior of the ratio between $\rho(\xi)$ and the rest energy density at infinity, $n_\infty mc^2$, is not monotonous in the temperature anymore, as is visible from the plots in the right panel of Fig. 3. As explained before, the reason for this is the smaller fraction of particles being accreted as the temperature goes up, leading to the difference between the behavior of n/n_∞ and $\rho/(nmc^2)$ (the first ratio decreases while the second one increases as temperature increases).

Next, we compute the radial and tangential pressures p_{rad} and p_{tan} of the gas. As already mentioned in the introduction, it was shown in [10] that at the horizon and in the low temperature limit $z \to \infty$ the relativistic kinetic gas behaves very differently than an isotropic perfect fluid, p_{tan} being almost an order of magnitude larger than p_{rad}. In the following, we analyze the behavior of the pressures as a function of ξ for finite temperatures. The radial and tangential pressures are obtained from the eigenvalues of $T^\mu{}_\nu$ corresponding to the spacelike eigenvectors; more specifically,

$$T^\mu{}_\nu e_r^\nu = p_{rad} e_r^\mu, \qquad T^\mu{}_\nu e_\vartheta^\nu = p_{tan} e_\vartheta^\mu,$$

with $\mathbf{e}_r$ the unit outgoing radial eigenvector of $T^\mu{}_\nu$ and $\mathbf{e}_\vartheta := r^{-1}\partial_\vartheta$. In Fig. 4 we show the behavior of these quantities divided by the rest mass energy nmc^2. For large values of ξ, the ratios $p_{rad}/(nmc^2)$ and $p_{tan}/(nmc^2)$ are approximately equal to each other, and they only depend on the temperature, consistent with an isotropic gas satisfying the ideal gas equation of state. However, as one approaches the horizon, $p_{tan}/(nmc^2)$ increases steadily while $p_{rad}/(mc^2)$

IOP Conf. Series: Journal of Physics: Conf. Series **831** (2017) 012009 doi:10.1088/1742-6596/831/1/012009

decreases close to the horizon, resulting in an anisotropic distribution of the pressure and a departure from the isotropic perfect fluid case. The ratio between p_{tan} and p_{rad} at the horizon varies between a factor of about 5 for $z = 1$ and about 10 for large z, see Fig. 5 and Ref. [10] for the case $z \to \infty$.

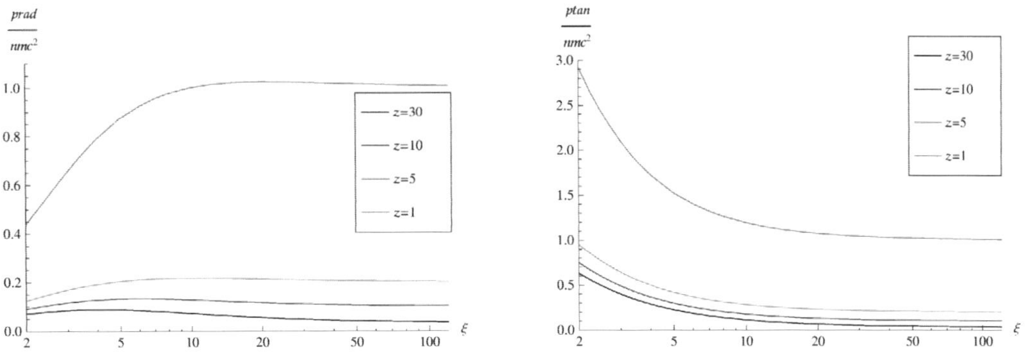

Figure 4. The ratio between the radial (left panel) and tangential (right panel) pressures and the rest energy density as a function of ξ for different values of the temperature.

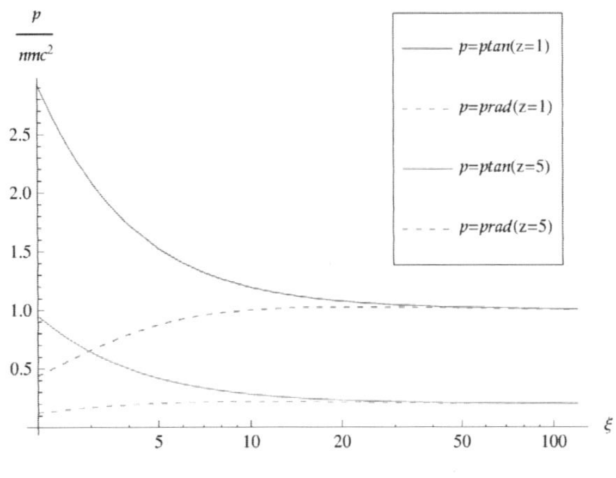

Figure 5. A comparison between the radial and tangential pressure profiles for two different values of z. The solid lines refer to the tangential pressure, the dashed lines to the radial one. While the ratio p_{tan}/p_{rad} converges to one as $\xi \to \infty$, it increases as one approaches the horizon, reaching a value of about 5 for $z = 1$ and about 8 for $z = 5$.

As discussed in [10], apart from the difference between the radial and tangential pressures, there is another property of the kinetic gas that distinguishes it from an isotropic perfect fluid configuration. In the latter, the current density is proportional to the four-velocity of the fluid which is also a timelike eigenvector of $T^{\mu}{}_{\nu}$. However, in the former case, $\mathbf{J} = J^{\mu}\partial_{\mu}$ does not coincide with the timelike eigenvector of $T^{\mu}{}_{\nu}$ in general. To quantify this difference, we introduce the hyperbolic angle θ defined by

$$\cosh\theta = -g(\mathbf{e}_0, \mathbf{u}), \qquad \sinh\theta = g(\mathbf{e}_r, \mathbf{u}),$$

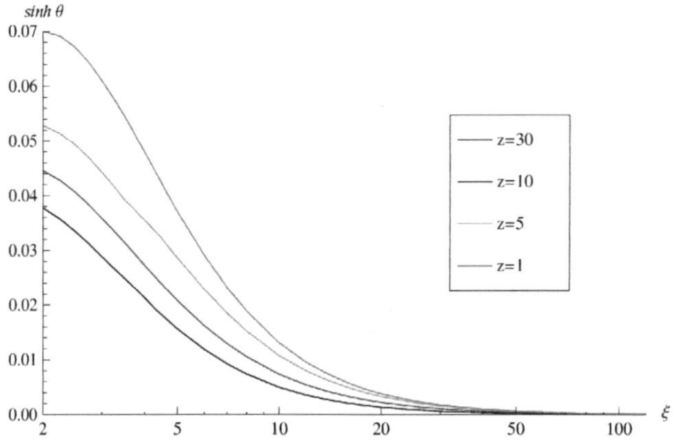

Figure 6. The quantity $\sinh\theta$ as a function of ξ for different values of the temperature. While $\sinh\theta \to 0$ as $\xi \to \infty$ for all values of z, $\sinh\theta$ is clearly different from zero close to the horizon, implying that the kinetic gas does not behave as a perfect fluid in the vicinity of the horizon.

where $\mathbf{u} := \mathbf{J}/n$ is the mean four-velocity of the particles. In the perfect fluid case, $\mathbf{u}$ coincides with $\mathbf{e}_0$ and $\theta = 0$; hence the angle θ measures the deviation from the perfect fluid case. In Fig. 6 we plot $\sinh\theta$ as a function of ξ for different temperatures. While asymptotically for $\xi \to \infty$ this quantity converges to zero for all values of z, it increases steadily as one approaches the horizon. In the limit $z \to \infty$, the value of $\sinh\theta$ at the horizon can be computed from the results in Section V.A of Ref. [10], giving $\sinh\theta \simeq 0.0345$. As the temperature increases, this value increases until it reaches $\sinh\theta \simeq 0.07$ when $z = 1$.

4. Comparison with the isotropic perfect fluid case

In this section, we wish to address the question of why our relativistic kinetic gas model behaves so much differently than the hydrodynamic model of Bondi-Michel accretion, even though in both cases the gas behaves exactly as an isotropic perfect fluid at infinity. Naively, one might have expected that in some limit the kinetic model should have reproduced the hydrodynamic flow, since the hydrodynamic equations can be obtained as an appropriate limit of a kinetic gas which is in (local) thermodynamic equilibrium [19, 18, 2]. Although in our model we have neglected collisions, one could still argue that our distribution function in Eq. (6) corresponds to an equilibrium distribution function and hence, the corresponding observables should be fluid-like. So why is the accretion flow in our model different than in the hydrodynamic case?

The answer comes from the observation that although, formally, the distribution function described by Eq. (6) looks like an equilibrium distribution function, in fact it is not. The reason for this can be traced back to the curved geometry of the Schwarzschild background and its causal properties, and relies on the fact that in our model in [10] the support of the distribution function is restricted to the region Γ_{accr} of phase space corresponding to particle trajectories originating from the reservoir. On the one hand, this means that only particles with large enough energies $E > mc^2$ are admitted in the gas. Although this eliminates the possibility of occupying bounded trajectories, it should be noted that the contributions from such trajectories would not affect the observables in the vicinity of the horizon, since they are confined to the region $r > 3r_H/2$ outside the photon sphere. On the other hand, the restriction to Γ_{accr} also eliminates the possibility of occupying those particle trajectories that emanate from the white hole (see Fig. 7) and whose contributions would affect the observables close to the horizon. Therefore, since it is truncated to the subset Γ_{accr}, the distribution function (6) is *not* an equilibrium distribution function.

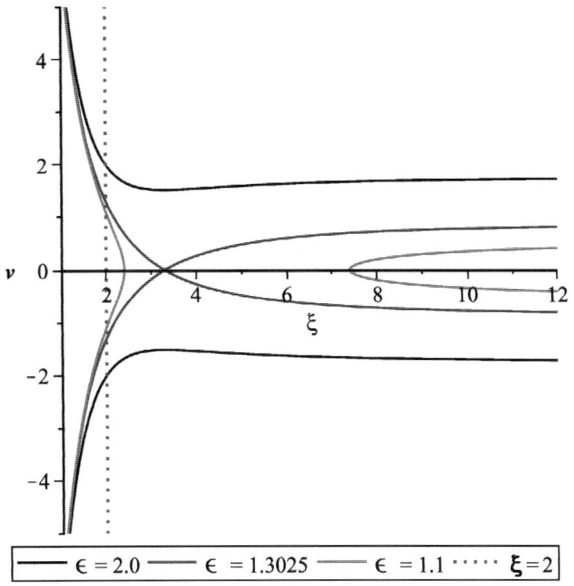

Figure 7. (adapted from Fig. 2 in [10]) Projection of the phase diagram onto to $(\xi, v = \dot{r}/mc)$-plane illustrating the particle trajectories for different energy levels ε and total dimensionless angular momentum $\lambda = 6$. The black curve in the region $v < 0$ corresponds to an incoming particle from infinity which is absorbed by the black hole, while the black curve in the region $v > 0$ describes an outgoing particle that is emitted from the white hole and escapes to infinity. The red curve in the region $\xi > 6$ describes a particle that is incoming from infinity but has large enough angular momentum $\lambda > \lambda_c(\varepsilon)$ to be reflected at the potential barrier, and the red curve in the region $\xi < 3$ describes a particle that is emitted by the white hole, is reflected at the potential barrier and absorbed by the black hole. The blue curve describes the separatrix and corresponds to the energy level ε such that $\lambda_c(\varepsilon) = \lambda$.

In the collisionless case, the restriction to Γ_{accr} is appropriate, since if a particle trajectory is occupied at one instant, it is occupied for all times, such that eliminating the particle trajectories emanating from the white hole makes sense from a physical point of view. In fact, if one considers the distribution function (6) on the full phase space instead of restricting it to Γ_{accr}, one obtains an isotropic perfect fluid configuration with four-velocity u parallel to the Killing vector field $k = \partial_t$. However, this configuration describes a static flow which fails to be regular at the horizon, and thus it does not describe a Bondi-Michel-like flow or any other physically well-defined solution.

Clearly, the situation changes when collisions are taken into account, since in this case a binary collision may cause an unoccupied trajectory to become occupied after the collision takes place. Hence, in this case, it does not make sense to restrict the phase space to Γ_{accr}, and the bounded trajectories and those emanating from the white hole have to be taken into account. We intend to analyze the effects of collisions in future work.

5. Conclusions

In this article, extending previous work [10], we have presented a detailed analysis of the properties of the spherical, steady-state accretion of a relativistic, collisionless kinetic gas into a Schwarzschild black hole. Assuming that in the asymptotic region the gas is isotropic and described by an equilibrium distribution function, we have computed the relevant physical quantities such as the accretion rate, particle density, energy density and radial and tangential pressures, and we have analyzed their behavior as a function of the temperature of the gas at infinity and/or the radial coordinate. In particular, we have shown that for given mass of the black hole and given particle density number at infinity, the energy accretion rate diverges when $z \to 0$ or $z \to \infty$, having a minimum when the ratio $z = mc^2/(k_B T)$ between the rest mass of the particles and their thermal energy is of the order one. We attributed this behavior to the competition between two different effects: the smaller fraction of particles being accreted by the black hole versus their higher internal energy as the temperature increases.

Further, we have analyzed the difference between the properties of the collisionless flow to those of an isotropic perfect fluid, as is assumed in the Bondi-Michel model. This difference manifests itself in at least three different ways. First, in the low temperature limit, the accretion rate in the collisionless case is much lower than in the fluid case, leading to mass accretion rates of the order of $10^{-21} M_\odot/yr$ for accretion of the ionized component of the interstellar medium by a stellar mass black hole of mass $10 M_\odot$ under typical conditions [13]. Second, in the collisionless case, although equal at infinity, the radial and tangential pressures differ from each other at finite radius (the difference being largest at the horizon) implying that the gas is anisotropic. Third, in the collisionless case, the current density does not necessarily agree with the timelike eigenvector of the stress energy-momentum tensor. This difference can be quantified by an hyperbolic angle θ whose behavior we have analyzed. This behavior shows that in the vicinity of the horizon, the collisionless kinetic gas is different from a perfect fluid.

Finally, we have provided an explanation for the fact that in our model the kinetic gas does not behave as an isotropic perfect fluid, although it does so at infinity. As we show, the origin for this effect is geometrical, the non-trivial causal structure of the Schwarzschild black hole leading to a discrimination between the particle trajectories that end at the black hole and those that emanate from the white hole. It should be interesting to investigate the effects introduced by collisions, and to check whether such effects diminish the differences reported in this article between the kinetic and isotropic perfect fluid models.

Acknowledgement

We thank Darío Núñez and Thomas Zannias for fruitful discussions and questions that motivated some of the discussions presented in this work. This investigation was partially supported by CONACyT Grants No. 577742 and No. 271904, and by a CIC Grant to Universidad Michoacana.

References

[1] Cercignani C and Kremer G 2002 *The Relativistic Boltzmann Equation: Theory and Applications* (Basel: Birkhäuser)
[2] Stewart J 1971 *Non-Equilibrium Relativistic Kinetic Theory* (Berlin: Springer-Verlag)
[3] Ehlers J 1971 General relativity and kinetic theory *General Relativity and Cosmology* ed Sachs R (New York: Acedemic) pp 1–70
[4] Andréasson H 2011 *Living Reviews in Relativity* **14** URL http://www.livingreviews.org/lrr-2011-4
[5] Lindquist R 1966 *Ann. Phys. (N.Y.)* **37** 487–518
[6] Ehlers J 1973 Survey of general relativity theory *Relativity, Astrophysics and Cosmology* ed Israel W (Dordrecht: D. Reidel) pp 1–125
[7] Sarbach O and Zannias T 2013 *AIP Conf. Proc.* **1548** 134–155
[8] Sarbach O and Zannias T 2014 *AIP Conf. Proc.* **1577** 192–207
[9] Sarbach O and Zannias T 2014 *Class. Quantum Grav.* **31** 085013
[10] Rioseco P and Sarbach O 2016 arXiv:1611.02389

[11] Bondi H 1952 *Monthly Notices Roy Astronom. Soc.* **112** 195–204
[12] Michel F 1972 *Astrophysics and Space Science* **15** 153–160
[13] Shapiro S and Teukolsky S 1983 *Black Holes, White Dwarfs, and Neutron Stars* (New York: John Wiley & Sons)
[14] Chaverra E and Sarbach O 2015 *Class. Quantum Grav.* **32** 155006
[15] Zel'dovich Y and Novikov I 1971 *Stars and Relativity* (Mineola, New York: Dover Publications, Inc.)
[16] F Jüttner 1911 *Annal. Phys.* **34** 856–882
[17] F Jüttner 1911 *Annal. Phys.* **35** 145–161
[18] Israel W 1963 *J. Math. Phys.* **4** 1163–1181
[19] Huang K 1987 *Statistical Mechanics* (New York, USA: John Wiley & Sons)

IOP Conf. Series: Journal of Physics: Conf. Series **831** (2017) 012010 doi:10.1088/1742-6596/831/1/012010

Relativistic Static Thin Disks of Polarized Matter

Anamaria Navarro and F. D. Lora-Clavijo and Guillermo A. González

Grupo de Investigación en Relatividad y Gravitación, Escuela de Física, Universidad Industrial de Santander, A. A. 678, Bucaramanga 680002, Colombia.

E-mail: AN: ana.navarro1@correo.uis.edu.co ; FDLC: fadulora@uis.edu.co; GAG: guilllermo.gonzalez@saber.uis.edu.co

Abstract. An infinite family of exact solutions of the electrovacuum Einstein-Maxwell equations is presented. The family is static, axially symmetric and describe thin disks composed by electrically polarized material in a conformastatic spacetime. The form of the conformastatic metric allows us to write down the metric functions and the electromagnetic potentials in terms of a solution of the Laplace equation. We find a general expression for the surface energy density of the disk, the pressure, the polarization vector, the electromagnetic fields and the velocity rotation for circular orbits. As an example, we present the first model of the family and show the behavior of the different physical variables.

1. Introduction

Exact explicit solutions of Einstein equations with axial symmetry have played a crucial role in astrophysical applications of general relativity. In particular, disk-like configurations have been used to model different kind of configurations like galactic disks, accretion disks around compact objects and also to explain the mechanism for extracting energy from the black hole to power the launching of the jet.

The first models of relativistic thin disks were obtained in 1968 by Bonnor and Sackfield [1] and in by 1969 Morgan and Morgan [2]. Since then, many authors have obtained different types of exact solutions of static thin disks [3–9], and stationary thin disks [10, 11], whereas the stability of this systems was studied in [12] using a first order perturbation on the energy-momentum tensor. On the other hand, due to the high relevance of electromagnetic fields in general relativity, there have been obtained many solutions that correspond to disks with electric and magnetic fields generated by electrically charged matter and electric currents, see for instance [13–22].

Electric or magnetic dipoles have not been studied very much in the context of exact solutions. The only related articles are [23, 24], where they analyze the electromagnetic fields generated by electric and magnetic dipole layers in a Schwarzschild spacetime. Since there have not been obtained exact solutions of the Einstein-Maxwell that may be interpreted as electric polarized source, in this paper we construct an infinite family of relativistic static thin disks composed by polarized material. In order to do so, we solve the Einstein-Maxwell equations in a conformastatic spacetime using the distributional approach of tensors. Finally, we obtain the general expressions for the surface energy-momentum tensor, the electric field, the polarization vector and analyze them for the first model of the family.

2. Einstein Maxwell Field Equations

The Einstein-Maxwell system of equations for a continuum media, in Heaviside-Lorentz geometrized units ($c = 8\pi G = \mu_0 = \epsilon = 1$, can be written as

$$G_{\alpha\beta} = T^M_{\alpha\beta} + T^F_{\alpha\beta} + T^{FM}_{\alpha\beta}, \qquad F^{\alpha\beta}{}_{;\beta} = M^{\alpha\beta}{}_{;\beta}, \tag{1}$$

where the components of the electromagnetic tensor $F_{\alpha\beta}$ and the components of the polarization tensor $M_{\alpha\beta}$ are given in terms of the electromagnetic potential A_α and the electric polarization vector P_α as

$$F_{\alpha\beta} = A_{\beta,\alpha} - A_{\alpha,\beta}, \qquad M^{\alpha\beta} = P^\alpha u^\beta - P^\beta u^\alpha. \tag{2}$$

Here u_α corresponds to the four velocity vector of the fluid. As we can see from equations (1), the energy momentum tensor is splitted into three terms [25]

$$T_{\alpha\beta} = T^M_{\alpha\beta} + T^F_{\alpha\beta} + T^{FM}_{\alpha\beta}, \tag{3}$$

$$T^F_{\alpha\beta} = F_{\alpha\mu}F_\beta{}^\mu - \frac{1}{4}g_{\alpha\beta}F_{\mu\nu}F^{\mu\nu}, \tag{4}$$

$$T^{FM}_{\alpha\beta} = -F_{\alpha\mu}M_\beta{}^\mu, \tag{5}$$

which correspond to the matter $T^M_{\alpha\beta}$, the electromagnetic fields $T^F_{\alpha\beta}$ and the electromagnetic interaction with the polarized matter $T^{FM}_{\alpha\beta}$, respectively. It is worth mentioning that the matter energy momentum tensor, the electromagnetic four potential and the electric polarization vector are calculated by using the displace, cut and reflection inverse method [26].

3. Axially Symmetric Solutions for Electrically Polarized Disks

In order to write down the Einstein-Maxwell field equations for electrically polarized infinitesimal thin disks, the line element for a conformastatic axially symmetric space-time in cylindrical coordinates $x^a = (t, r, \phi, z)$, can be written as [27]

$$ds^2 = -e^{2\psi}dt^2 + e^{-2\psi}\left(dr^2 + r^2d\phi^2 + dz^2\right), \tag{6}$$

where the metric function ψ depends on the coordinates (r, z). The fact that the solution is not magnetized, the electromagnetic four potential takes the form $A_\alpha = (-\chi, 0, 0, 0)$, where the new function χ also depends on the coordinates (r, z). Based on these, the non-zero component of the polarization vector is given by

$$P_r = e^{-\psi}\Pi_{rt}, \tag{7}$$

where $\Pi_{\alpha\beta}$ are the components of the polarization tensor on the disk, which are related with the components of $M^{\alpha\beta}$ through the relation $M^{\alpha\beta} = \Pi^{\alpha\beta}\delta(z)$, where $\delta(z)$ is the Dirac distribution with support in the surface $z = 0$. As we can see, the antisymmetry of the polarization tensor implies that $\Pi_{rt} = -\Pi_{tr} = P_r e^\psi$.

Solving the Einstein-Maxwell equations (1), outside and inside of the disk, we obtain that its surface energy density σ, pressure (p_r, p_ϕ, p_z) and electric components $(E_{(r)}, E_{(z)})$ are

$$\begin{aligned} &\sigma = e^\psi\left(4\psi_{,z} + \sqrt{2}\psi_{,r}P_r\right), \quad p_r = -\sqrt{2}\psi_{,r}e^\psi P_r, \quad p_\phi = 0 \quad p_z = 0, \\ &E_{(r)} = -\sqrt{2}\, e^\psi\psi_{,r}, \quad E_{(z)} = -\sqrt{2}\, e^\psi\psi_{,z}, \end{aligned} \tag{8}$$

with

$$P_r = -\frac{2\sqrt{2}e^\psi}{r}\int r\psi_{,z}e^{-\psi}\mathrm{d}r, \tag{9}$$

where it was found that the electric potential χ can be written as $\chi = \sqrt{2}\left(e^{\psi} - 1\right)$ and the metric function ψ satisfies the following equation

$$\nabla^2\psi - \nabla\psi \cdot \nabla\psi = 0, \tag{10}$$

which can be cast as the Laplace equation $\nabla^2\left(e^{-\psi}\right) = 0$. Furthermore, as we requiere the space-time to be asymptotically flat, the function $e^{-\psi}$ must be 1 at infinity, so we can write

$$e^{-\psi} = 1 - U\,, \tag{11}$$

where U is any solution of the Laplace equation that vanishes at infinity. Accordingly, equation (11) allows us to write the physical quantities of the disk in terms of the function U, so we have

$$\begin{aligned}
\sigma &= \frac{4}{(1-U)^2}\left[U_{,z} - \frac{U_{,r}}{(1-U)r}\int_0^\infty U_{,z} r\mathrm{d}r\right], \\
p_r &= \frac{4U_{,r}}{r(1-U)^3}\int_0^\infty U_{,z} r\mathrm{d}r\,, \qquad P_{(r)} = \frac{2\sqrt{2}}{(1-U)^2 r}\int_0^\infty U_{,z} r\mathrm{d}r\,, \\
E_{(r)} &= -\frac{\sqrt{2}U_{,r}}{(1-U)^2}\,, \qquad E_{(z)} = -\frac{\sqrt{2}U_{,z}}{(1-U)^2}\,.
\end{aligned} \tag{12}$$

It can be seen that, the physical features of the disk depend entirely on the election of the solution of Laplace equation U.

4. A Particular Family of Polarized Disks

Considering the axial symmetry of the system, we take as solution of Laplace equation the following function

$$U_n = -\sum_{l=0}^{n} \frac{C_l P_l\left(z/R\right)}{R^{l+1}}\,, \tag{13}$$

with C_l constants, $P_l(\cos\theta)$ the Legendre polynomials and $R = \sqrt{r^2 + z^2}$. As we can see, this solution is zero once R tends to infinity, which guarantees together with (11) that the spacetime is asymptotically flat. Now, in order to introduce the discontinuities in the metric tensor and in the electromagnetic four potential, we apply the displace, cut and reflect method given by the the transformation

$$z \to |z| + a\,, \tag{14}$$

being a a positive constant, to the function U_n. Accordingly, we now have

$$U_n = -\sum_{l=0}^{n} \frac{C_l P_l\left(\left(|z| + a\right)/R\right)}{R^{l+1}}\,, \tag{15}$$

with $R = \sqrt{r^2 + (|z| + a)^2}$. Hence, the general expressions for the physical features of the disk are given by

$$\sigma = \frac{4\sum\limits_{l=0}^{n}\xi_1}{\left(1 + \sum\limits_{l=0}^{n}\xi_4\right)^2} - \frac{4\sum\limits_{m=0}^{n}\xi_2 \int\limits_0^\infty \sum\limits_{l=0}^{n}\xi_3 r\mathrm{d}r}{\left(1 + \sum\limits_{l=0}^{n}\xi_4\right)^3}\,,$$

$$p_r = \frac{\sum_{m=0}^{n} \xi_2 \int_0^{\infty} \sum_{l=0}^{n} \xi_3 r dr}{\left(1+\sum_{l=0}^{n}\xi_4\right)^3} \qquad P_{(r)} = -\frac{2\sqrt{2}\int_0^{\infty} \sum_{l=0}^{n} \xi_3 r dr}{\left(1+\sum_{l=0}^{n}\xi_4\right)^2 r}, \tag{16}$$

$$E_{(r)} = -\frac{\sqrt{2}r\sum_{l=0}^{n}\xi_5}{\left(1+\sum_{l=0}^{n}\xi_4\right)^2}, \qquad E_{(z)} = -\frac{\sqrt{2}\sum_{l=0}^{n}\xi_1}{\left(1+\sum_{l=0}^{n}\xi_4\right)^2},$$

with

$$\xi_1 = C_l(l+1)P_{l+1}(a/R_0)/R_0^{(l+2)}, \quad \xi_2 = C_m P'_{m+1}(a/R_0)/R_0^{(m+3)},$$

$$\xi_3 = C_l(l+1)P_{l+1}(a/R_0)/R_0^{(l+2)}, \quad \xi_4 = C_l P_l(a/R_0)/R_0^{l+1}, \quad \xi_5 = C_l P'_l(a/R_0)/R_0^{(l+3)},$$

and $R_0 = \sqrt{r^2+a^2}$. In order to obtain a particular solution, it is necessary to specify the value n and the constants C_l.

As an example of electrically polarized thin disks, we consider the solution of the Laplace equation corresponding to the first term in the equation (13), that is, when $n = 0$

$$U_0 = -\frac{C_0}{R}, \tag{17}$$

with $R = \sqrt{r^2 + (|z|+a)^2}$. In consequence the a-dimensional surface energy density $\bar{\sigma}$, the radial pressure $\bar{p}_r$ and the radial component of the polarization vector $\bar{P}_{(r)}$ can be written as

$$\bar{\sigma} = \frac{\sigma}{\sigma_0} = \left(\frac{1+\bar{C}_0}{\bar{R}_0+\bar{C}_0}\right)^3, \quad \bar{p}_r = \frac{p_r}{p_{r_0}} = \frac{1}{\bar{R}_0}\left(\frac{1+\bar{C}_0}{\bar{R}_0+\bar{C}_0}\right)^3, \quad \bar{P}_{(r)} = \frac{-2\sqrt{2}\bar{C}_0\bar{R}_0}{(\bar{R}_0+\bar{C}_0)^2\bar{r}}, \tag{18}$$

where the quantities $r = a\bar{r}$ and $C_0 = a\bar{C}_0$, $R_0 = a\bar{R}_0$ were introduced in order to dimensionless the state variables of the system. Here $\sigma_0 = \sigma|_{(r=0)}$, $p_{r_0} = p_r|_{(r=0)}$. Moreover, if we demand σ and p_r to satisfy the energy conditions [28–30], in order to obtain well behaved solutions, the value of the constant C_0 must be $C_0 = a$, which implies that $\bar{C}_0 = 1$.

On the other hand, the electric potential and the components of the electric field are given by the following expressions

$$\chi = \frac{-\sqrt{2}C_0}{R+C_0}, \qquad E_{(r)} = -\frac{2rC_0}{R(R+C_0)^2}, \qquad E_{(z)} = -\frac{2zC_0}{R(R+C_0)^2}. \tag{19}$$

As we can see, if we even the electrical potential χ to a constant Γ and solve for $\bar{r}$, we obtain an equation for the equipotential lines

$$\bar{r} = \pm\sqrt{\nu - (|\bar{z}|+1)^2}, \qquad \nu = a^2\bar{C}_0\left(\frac{\sqrt{2}}{\Gamma+1}\right), \tag{20}$$

where $z = a\bar{z}$. In the same way, in order to get the electric field lines, we replace the expressions for $E_{(r)}$ and $E_{(z)}$ into the following equation

$$\frac{dz}{E_{(z)}} = \frac{dr}{E_{(r)}},$$

IOP Conf. Series: Journal of Physics: Conf. Series **831** (2017) 012010 doi:10.1088/1742-6596/831/1/012010

and solve again for $\bar{r}$

$$\bar{r} = \omega(|\bar{z}| + 1), \qquad \omega = \sqrt{2\left(\frac{\bar{C}_0}{\bar{\kappa}}\right)^2 - 1}, \tag{21}$$

where $\kappa = a\bar{\kappa}$.

In figure 1, we show the dimensionless profiles for the surface energy density $\bar{\sigma}$, radial pressure $\bar{p}_r$ (left panel), radial polarization $\bar{P}_r$ (middle panel), the electric field and the equipotential lines (right panel) as a function of $\bar{r}$ for $\bar{C}_0 = 1$. From these plots, it can be observed that $\bar{\sigma}$ and $\bar{p}_r$ are everywhere positives, the maximum values occur at the center of the disk and when $\bar{r}$ increases they decrease. On the other hand, the electric polarization vector $\bar{P}_r$ tends to zero when $\bar{r}$ tends to infinity and has a singularity at the center of the disk due to the form of function U_0, which is a solution of the Laplace equation.

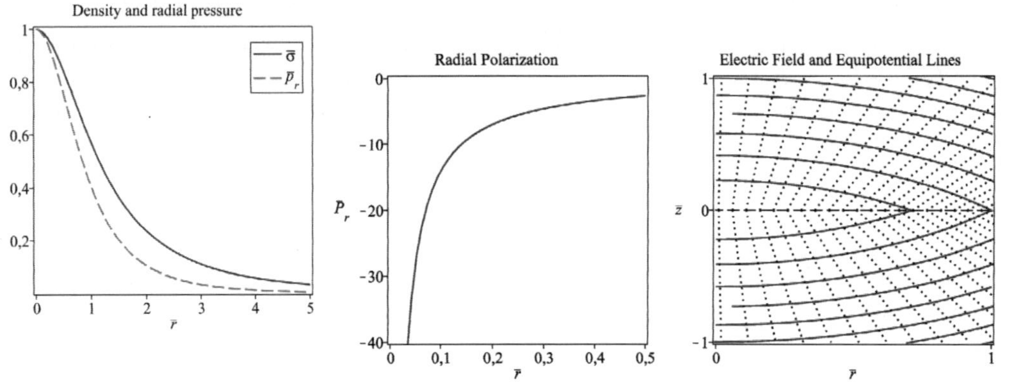

Figure 1. Physical features of the polarized disk $n = 0$ with $\bar{C}_0 = 1$.

5. ROTATION CURVES

From the astrophysical point of view, an important aspect that can be studied in this system is the circular geodesics along the plane of the disk ($z = 0$), in order to qualitatively determine its resemblance with the observed rotation curves. For the line element (6) and assuming r constant, such movement has a velocity given by

$$u^\alpha = u^t\,(1, 0, \Omega, 0)\ , \tag{22}$$

where $\Omega = u^\phi/u^t$ is the angular velocity. By imposing the normalization condition $g_{\alpha\beta}u^\alpha u^\beta = -1$, we obtain the relation

$$\left(u^t\right)^2 = \frac{e^{-2\psi}}{1 - V_c^2}, \tag{23}$$

where $V_c = e^{-2\psi} r\Omega$ is the circular o tangential velocity. On the other hand, from the geodesic equation

$$\frac{du_\alpha}{d\tau} = \frac{1}{2} g_{\mu\nu,\alpha} u^\mu u^\nu\ , \tag{24}$$

the angular velocity can be written as

$$\Omega = \frac{\psi_{,r}\, e^{4\psi}}{r(1 - r\psi_{,r})}, \tag{25}$$

and then we get the following expression for the circular velocity in terms of the function U

$$V_{nc} = \sqrt{\frac{rU_{n,r}}{1 - U_n - rU_{n,r}}}. \tag{26}$$

For the cases above mentioned, that is $n = 0$, the rotation curve is given as

$$V_{0c} = \bar{r}\sqrt{\frac{\bar{C}_0}{\bar{R}_0{}^3 + \bar{C}_0}}, \tag{27}$$

The rotation curve is displayed in figure 2, and its behavior is similar with the predicted rotation curves from Newtonian gravity for spiral galaxies [31].

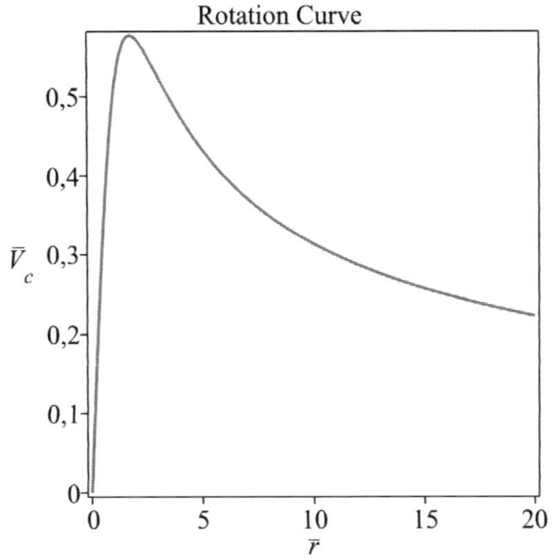

Figure 2. Rotation curves for $n = 0$

6. Conclusions

A infinite family of exact solutions of the Einstein-Maxwell equations was obtained for an axially symmetric conformastatic spacetime. The solution describes thin disks electrically polarized. This model was obtained by using the distributional approach of tensors, and introducing finite discontinuities in the first normal derivatives of the metric tensor and the electromagnetic four potential. We analyzed the first model of the family, that is, the model $n = 0$, finding that the disk behaves in a reasonable way, that is, the functions are everywhere positive, smooth, decay when the distance from the center increases and show no singularities except for the electrically polarization vector. Despite of the infinite extension of the disk the physical features are concentrated in the central region as we could expect from a physical system. This work serves as a starting point for the study of electromagnetic polarized matter within the exact solutions of disk-like configurations.

7. Acknowledgment
A. N wants to thanks the financial support from COLCIENCIAS and Universidad Industrial de Santander. F.D.L-C gratefully acknowledges the financial support from Vicerrectoría de Investigación y Extensión, Universidad Industrial de Santander, grant number 1822. G. A. G. was supported in part by VIE-UIS, under Grants No. 1347 and No. 1838, and by COLCIENCIAS, Colombia, under Grant No. 8840.

References
[1] Bonnor W B and Sackfield A 1968 *Commun. math. Phys.* **8** 338
[2] Morgan T and Morgan L 1969 *Phys. Rev.* **138** 1097
[3] Morgan T and Morgan L 1970 *Phys. Rev.* **D 2** 2756
[4] Voorhees B H 1972 *Phys. Rev.* **D 5** 2413
[5] Lynden-Bell D and Pineault S 1978 *Mon. Not. R. Astron. Soc.* **185** 679
[6] Letelier P S and Oliveira S R 1987 *J. Math. Phys.* **28** 165
[7] Lemos J P S 1989 *Class. Quantum Grav.* **6** 1219
[8] Bičák D, Lynden-Bell D and Katz J 1993 *Phys. Rev.* **D 47** 4334
[9] González G A and Pimentel O M 2016 *Phys. Rev.* **D93** 044034
[10] Lynden-Bell D y Pineault S 1978 *Mon. Not. R. Astron. Soc.* **185** 695
[11] Bičák J and Ledvinka T 1993 *Phys. Rev. Lett.* **71** 1669
[12] Ujevic M and Letelier P S 2004 *Phys. Rev.* **D 70** 084015
[13] García-Reyes G and González G A 2007 *J. Phys.* **37** 1094
[14] Lora-Clavijo F, Ospina-Henao P and Pedraza J 2010 *Phys. Rev.* **D 82** 084005
[15] García G and González G A 2004 *Phys. Rev.* **D 69** 124002
[16] Letelier P S 1999 *Phys. Rev.* **D 60** 104042
[17] Vogt D and Letelier P 2007 *Class. Quantum Grav.* **21** 3369
[18] Katz J, Bičák J and Lynden-Bell D 1999 *Class. Quantum Grav.* **16** 4023
[19] González G A, Gutiérrez-Piñeres A C and Ospina P 2008 *Phys. Rev.* **D 78** 064058
[20] Vogt D and Letelier P S 2004 *Phys. Rev.* **D 70** 064003
[21] García-Reyes G and González G A 2004 *Class. Quantum Grav.* **21** 4845
[22] García-Reyes G and González G A 2004 *Phys. Rev.* **D70** 104005
[23] Gürlebeck N, Bičák J and Gutiérrez-Piñeres A C 2011 *Phys. Rev.* **D 83** 124023
[24] Gürlebeck N, Bičák J and Gutiérrez-Piñeres A C 2011 *Gen. Rel. Grav.* **43** 3301
[25] Grot R A 1970 *J. Math. Phys.* **11** 109
[26] Kuzmin G 1956 *Astronomicheskii Zhurnal* **33** 27
[27] Synge J L 1966 *Relativity: The General Theory* (North–Holland Publishing Company)
[28] Poisson E 2004 *A Relativistit's Toolkit: The Mathematics of Black–Hole Mechanics* (Cambridge University Press)
[29] Pimentel O M, González G A and Lora-Clavijo F D 2016 *Gen. Rel. Grav.* **48** 124
[30] Pimentel O M, Lora-Clavijo F D and González G A 2016 (*Preprint* `arXiv:1612.03299 [gr-qc]`)
[31] Rubin V C and Ford W K 1970 *ApJ* **151** 379

IOP Conf. Series: Journal of Physics: Conf. Series **831** (2017) 012011 doi:10.1088/1742-6596/831/1/012011

An equivalent system of Einstein Equations

J. Ospino[1] and J.L. Hernández-Pastora[2], L.A. Núñez[3,4]

[1]Departamento de Matemática Aplicada and Instituto Universitario de Física Fundamental y Matemáticas, Universidad de Salamanca, Salamanca, Spain
[2]Escuela de Física, Universidad Industrial de Santander, Bucaramanga-Colombia
[3]Departamento de Física, Universidad de Los Andes, Mérida Venezuela.

E-mail: j.ospino@usal.es , jlhp@usal.es and lnunez@uis.edu.co

Abstract. We provide the full set of equations governing the evolution of self-gravitating systems by using a tetrad formalism in General Relativity and the orthogonal splitting of the Riemann tensor. We apply this formalism to spherical case and found that: the only static solution with homogeneous energy density, under the physical reasonable considerations, is the Schwarzchild solution; the conditions under which a non-dissipative fluid is static and finally that shear-free and isotropic pressure conditions are equivalent, for non-dissipative fluids with homogeneous energy density.

1. Introduction

Timelike congruence is one of the main subjects of General Relativity because matter particles follow timelike trajectories. Therefore, the study of geometrical and kinematic aspects of timelike curves is fundamental in the analysis of the evolution of self-gravitating fluids and there it is common the use of a framework based on the well known 1+3 formalism [1, 2, 3, 4, 5, 6, 7]. In this formalism, any tensor quantity can be split in its components along of a tangent vector to a timelike congruence $\mathbf{V}$, and the components totally orthogonal to it.

From the orthogonal splitting of the first covariant derivative of $\mathbf{V}$ into irreducible parts, we obtain the kinematical variables, the expansion rate Θ, the shear rate $\sigma_{\alpha\beta}$ and the rotation rate $\omega_{\alpha\beta}$. Next, from the Ricci identities we find the evolution for these variables, the constraint equations. Finally from the Bianchi identities we obtain the evolution and constraint equations for the electric and magnetic part of the Weyl tensor.

In general we have a system of six evolution and six constraint equations of first order; considered as a set of evolution equations for the $1 + 3$ covariant variables. This system is solved for a given the equation of state for the material. However, they do not form a complete set of equations guaranteeing the existence of a corresponding metric and connections. Therefor we need to use a tetrad description.

In this work we shall use a tetrad formalism to obtain two set of first-order differential equations:

1)The first one is obtained from Ricci identities, which can be used to define the physical variables and the scalars of Weyl tensor.

2) The second one is obtained from Bianchi identities and can be interpreted as evolution equations for the physical variables and the scalars of Weyl tensor.

Later, we use this two set equations to analyze the spherical case and find some new results.

IOP Conf. Series: Journal of Physics: Conf. Series **831** (2017) 012011 doi:10.1088/1742-6596/831/1/012011

The paper is organized as follows: In the next section we present the strategy and general formalism and write all relevant equations. Section 3 is devoted to the analysis of spherical case and we end with a short final remark for future works.

2. The strategy and general formalism

The strategy we shall follow is to compile two independent sets of equations, expressed in terms of scalar functions, which contain the same information than the Einstein system.

Let us also choose an orthogonal unitary tetrad:

$$e_\alpha^{(0)} = V_\alpha,\ e_\alpha^{(1)} = K_\alpha,\ e_\alpha^{(2)} = L_\alpha \text{ and } e_\alpha^{(3)} = S_\alpha. \tag{1}$$

As usual, $\eta_{(a)(b)} = g_{\alpha\beta}e^\alpha_{(a)}e^\beta_{(b)}$, with $a = 0,\ 1,\ 2,\ 3$, i.e. latin indices label different vectors of the tetrad. Thus, the tetrad satisfies the standard relations:

$$\begin{aligned} V_\alpha V^\alpha = -K_\alpha K^\alpha &= -L_\alpha L^\alpha = -S_\alpha S^\alpha = -1 \\ V_\alpha K^\alpha = V_\alpha L^\alpha = V_\alpha S^\alpha &= K_\alpha L^\alpha = K_\alpha S^\alpha = S_\alpha L^\alpha = 0 \end{aligned} \tag{2}$$

With the above tetrad (1) we shall also define the corresponding directional derivatives operators

$$f^\bullet = V^\alpha\partial_\alpha f; \quad f^\dagger = K^\alpha\partial_\alpha f \quad \text{and} \quad f^* = L^\alpha\partial_\alpha f. \tag{3}$$

The first set can be considered purely geometrical and emerges from the projection of the Riemann tensor along the tetrad [8], i.e.

$$\begin{aligned} 2V_{\alpha\,;[\beta;\gamma]} &= R_{\delta\alpha\beta\gamma}V^\delta\,, & 2K_{\alpha\,;[\beta;\gamma]} &= R_{\delta\alpha\beta\gamma}K^\delta\,, \\ 2L_{\alpha\,;[\beta;\gamma]} &= R_{\delta\alpha\beta\gamma}L^\delta \quad \text{and} & 2S_{\alpha\,;[\beta;\gamma]} &= R_{\delta\alpha\beta\gamma}S^\delta\,; \end{aligned} \tag{4}$$

where $e^{(a)}_{\alpha\,;\beta\gamma}$ are the second covariant derivatives of each tetrad (7) vector indicated with $a = 0, 1, 2, 3$.

The second set emerges from the Bianchi identities i.e.

$$R_{\alpha\beta[\gamma\delta\,;\mu]} = R_{\alpha\beta\gamma\delta\,;\mu} + R_{\alpha\beta\mu\gamma\,;\delta} + R_{\alpha\beta\delta\mu\,;\gamma} = 0\,. \tag{5}$$

3. Spherical Case

In this section we shall present the relevant equations, for the spherically symmetric, locally anisotropic, disipative, collapsing matter configuration, written in terms of the kinematical variables –the four acceleration a_α, the expansion scalar Θ, and the shear tensor σ– and the structure scalars, which are some scalars functions related to the splitting of the Riemann Tensor.

3.1. The tetrad, the source and kinematical variables

To illustrate the above procedure we shall restrict to a spherically symmetric line element given by

$$\mathrm{d}s^2 = -A^2\mathrm{d}t^2 + B^2\mathrm{d}r^2 + R^2(\mathrm{d}\theta^2 + \sin^2(\theta)\mathrm{d}\phi^2)\,, \tag{6}$$

where the coordinates are: $x^0 = t$, $x^1 = r$, $x^2 = \theta$, and $x^3 = \phi$; with $A(t,r)$, $B(t,r)$ and $R(t,r)$ functions of their arguments.

For this case the tetrad can written as:

$$\begin{aligned} V_\alpha &= (-A,0,0,0) \qquad K_\alpha = (0,B,0,0) \\ L_\alpha &= (0,0,R,0) \quad \text{and} \quad S_\alpha = (0,0,0,R\sin\theta)\,. \end{aligned} \tag{7}$$

The covariant derivatives of the orthonormal tetrad are:

$$\begin{aligned} V_{\alpha;\beta} &= -a_1 K_\alpha V_\beta + \sigma_1 K_\alpha K_\beta + \sigma_2 (L_\alpha L_\beta + S_\alpha S_\beta), \\ K_{\alpha;\beta} &= -a_1 V_\alpha V_\beta + \sigma_1 V_\alpha K_\beta + J_1 (L_\alpha L_\beta + S_\alpha S_\beta), \\ L_{\alpha;\beta} &= \sigma_2 V_\alpha L_\beta - J_1 K_\alpha L_\beta + J_2 S_\alpha S_\beta \quad \text{and} \\ S_{\alpha;\beta} &= \sigma_2 V_\alpha S_\beta - J_1 K_\alpha S_\beta - J_2 L_\alpha S_\beta\,; \end{aligned} \tag{8}$$

where J_1, J_2, σ_1, σ_2 and a_1 are expressed in terms of the metric functions and their derivatives as:

$$\begin{aligned} J_1 &= \frac{1}{B}\frac{R'}{R}, & J_2 &= \frac{1}{R}\cot\theta, \\ \sigma_1 &= \frac{1}{A}\frac{\dot{B}}{B}, & \sigma_2 &= \frac{1}{A}\frac{\dot{R}}{R} \\ \text{and} \quad a_1 &= \frac{1}{B}\frac{A'}{A}; \end{aligned} \tag{9}$$

with primes and dots representing radial and time derivatives, respectively.

As we mentioned before we shall assume our source as a bounded, spherically symmetric, locally anisotropic, dissipative, collapsing matter configuration, described by a general energy momentum tensor, written in the "canonical" form, as

$$T_{\alpha\beta} = (\rho + P) V_\alpha V_\beta + P g_{\alpha\beta} + \Pi_{\alpha\beta} + \mathcal{F}_\alpha V_\beta + \mathcal{F}_\beta V_\alpha. \tag{10}$$

It is immediate to see that the physical variables can defined –at the Eckart frame where fluid elements are at rest– as:

$$\begin{aligned} \rho &= T_{\alpha\beta} V^\alpha V^\beta, & \mathcal{F}_\alpha &= -\rho V_\alpha - T_{\alpha\beta} V^\beta, \\ P &= \frac{1}{3} h^{\alpha\beta} T_{\alpha\beta}, & \text{and} \quad \Pi_{\alpha\beta} &= h^\mu_\alpha h^\nu_\beta \left(T_{\mu\nu} - P h_{\mu\nu} \right); \end{aligned} \tag{11}$$

with $h_{\mu\nu} = g_{\mu\nu} + V_\nu V_\mu$.

Observe that from the condition $\mathcal{F}^\mu V_\mu = 0$, and the symmetry of the problem, Einstein equations imply $T_{03} = 0$, thus

$$\mathcal{F}_\mu = \mathcal{F} K_\mu \quad \Leftrightarrow \quad \mathcal{F}_\mu = \left(0, \frac{\mathcal{F}}{B}, 0, 0 \right). \tag{12}$$

Clearly ρ is the energy density (the eigenvalue of $T_{\alpha\beta}$ for eigenvector V^α), $\mathcal{F}_\alpha$ represents the energy flux four vector; P corresponds to the isotropic pressure, and $\Pi_{\alpha\beta}$ is the anisotropic tensor, which can be expressed as

$$\Pi_{\alpha\beta} = \Pi_1 \left(K_\alpha K_\beta - \frac{h_{\alpha\beta}}{3} \right), \tag{13}$$

with

$$\Pi_1 = \left(2 K^\alpha K^\beta + L^\alpha L^\beta \right) T_{\alpha\beta}. \tag{14}$$

Finally, we shall express the kinematical variables –the four acceleration, the expansion scalar and the shear tensor– for a self-gravitating fluid as:

$$a_\alpha = V^\beta V_{\alpha;\beta} = aK_\alpha = \left(0, \frac{A'}{A}, 0, 0\right), \tag{15}$$

$$\Theta = V^\alpha_{;\alpha} = \frac{1}{A}\left(2\frac{\dot{B}}{B} + \frac{\dot{R}}{R}\right), \tag{16}$$

$$\sigma = \frac{1}{A}\left(\frac{\dot{B}}{B} - \frac{\dot{R}}{R}\right), \tag{17}$$

3.2. The orthogonal splitting of Riemann tensor and structure scalars
In this section we shall introduce a set of scalar functions, the structure scalars, obtained from the orthogonal splitting of the Riemann tensor (see [9, 10, 11]) which has proven to be very useful to express Einstein Equations.

Following [9], we can express the splitting of the Riemann tensor as

$$\begin{aligned} R_{\alpha\beta\mu\nu} &= 2V_\mu V_{[\alpha}Y_{\beta]\nu} + 2h_{\alpha[\nu}X_{\mu]\beta} + 2V_\nu V_{[\beta}Y_{\alpha]\mu} \\ &+ h_{\beta\nu}(\mathcal{X}_0 h_{\alpha\mu} - X_{\alpha\mu}) + h_{\beta\mu}(X_{\alpha\nu} - \mathcal{X}_0 h_{\alpha\nu}) \\ &+ 2V_{[\nu}Z^\gamma_{\ \mu]}\varepsilon_{\alpha\beta\gamma} + 2V_{[\beta}Z^\gamma_{\ \alpha]}\varepsilon_{\mu\nu\gamma} \end{aligned} \tag{18}$$

and its corresponding Ricci contraction as

$$\begin{aligned} R_{\alpha\mu} &= \mathcal{Y}_0 V_\alpha V_\mu - X_{\alpha\mu} - Y_{\alpha\mu} + \mathcal{X}_0 h_{\alpha\mu} + Z^{\nu\beta}\varepsilon_{\mu\nu\beta}V_\alpha \\ &+V_\mu Z^{\nu\beta}\varepsilon_{\alpha\nu\beta}\,; \end{aligned} \tag{19}$$

where the quantities $Y_{\alpha\beta}$, $X_{\alpha\beta}$ and $Z_{\alpha\beta}$ can be expressed as

$$Y_{\alpha\beta} = \frac{1}{3}\mathcal{Y}_0 h_{\alpha\beta} + \mathcal{Y}_1(K_\alpha K_\beta - \frac{1}{3}h_{\alpha\beta}), \tag{20}$$

$$X_{\alpha\beta} = \frac{1}{3}\mathcal{X}_0 h_{\alpha\beta} + \mathcal{X}_1(K_\alpha K_\beta - \frac{1}{3}h_{\alpha\beta}) \tag{21}$$

and

$$Z_{\alpha\beta} = Z(L_\alpha S_\beta - L_\beta S_\alpha) \tag{22}$$

with,

$$\mathcal{Y}_0 = 4\pi(\rho + 3P)\,, \qquad \mathcal{Y}_1 = \mathcal{E}_1 - 4\pi\Pi_1\,, \tag{23}$$

$$\mathcal{X}_0 = 8\pi\rho\,, \qquad \mathcal{X}_1 = -(\mathcal{E}_1 + 4\pi\Pi_1)\,, \tag{24}$$

$$\text{and} \qquad Z = 4\pi\mathcal{F}\,, \tag{25}$$

and the electric part of the Weyl tensor is written as

$$E_{\alpha\beta} = C_{\alpha\nu\beta\delta}V^\nu V^\delta = \mathcal{E}_1(K_\alpha K_\beta - \frac{1}{3}h_{\alpha\beta})\,. \tag{26}$$

3.3. Projections of Riemann Tensor
From the above system (4), –by using the covariant derivative of equations (8) and the projections of the orthogonal splitting of the Riemann tensor– we can obtain first set of

independent equations, for the spherical case, in terms of the J_1, J_2, σ_1, σ_2, a_1, defined in (9), and their directional derivatives , i.e.

$$\begin{aligned}
\sigma_1^{\bullet} - a_1^{\dagger} - a_1^2 + \sigma_1^2 &= -\frac{1}{3}(Y_0 + 2Y_1)\,, && (27)\\
\sigma_2^{\bullet} + \sigma_2^2 - a_1 J_1 &= \frac{1}{3}(Y_1 - Y_0)\,, && (28)\\
\sigma_2^{\dagger} + J_1(\sigma_2 - \sigma_1) &= -Z\,, && (29)\\
J_1^{\bullet} + J_1\sigma_2 - a_1\sigma_2 &= -Z\,, && (30)\\
J_1^{\dagger} + J_1^2 - \sigma_1\sigma_2 &= \frac{1}{3}(X_1 - X_0)\,, && (31)\\
J_2^{\bullet} + J_2\sigma_2 &= 0\,, && (32)\\
J_2^{\dagger} + J_1 J_2 &= 0 \quad \text{and} && (33)\\
J_1^2 - \frac{1}{R^2} - \sigma_2^2 &= -\frac{1}{3}(X_0 + 2X_1)\,. && (34)
\end{aligned}$$

3.4. Equations from Bianchi identities
The second set of equations for the spherical case, emerges from the independent Bianchi identities (5), and can be written as

$$\begin{aligned}
&a_1(-X_0 + X_1 - Y_0 + Y_1) + 3J_1Y_1 + 3Z^{\bullet}\\
&\qquad +6Z\sigma_1 + 3Z\sigma_2 - Y_0^{\dagger} + Y_1^{\dagger} = 0 && (35)\\
&X_0^{\bullet} - X_1^{\bullet} - 6a_1Z - 3J_1Z + (Y_0 - Y_1 - X_1)\,\sigma_1\\
&+ (Y_0 + 2Y_1 - X_1)\,\sigma_2 + X_0(\sigma_1 + \sigma_2) - 3Z^{\dagger} = 0 && (36)\\
&X_0^{\bullet} + 2X_1^{\bullet} + 2X_0\sigma_2 - 6J_1Z\\
&\qquad + (4X_1 + 2Y_0 - 2Y_1)\,\sigma_2 = 0 && (37)\\
&X_0^{\dagger} + 2X_1^{\dagger} + 6J_1X_1 + 6Z\sigma_2 = 0\,. && (38)
\end{aligned}$$

3.5. Spherical Cases of Study
3.5.1. The static case In the line element (6) we can assume, without any loss of generality, $R = r$ and integrate (28) to obtain

$$A = C_1 e^{\int B^2 r(Y_0 - Y_1)dr} \tag{39}$$

and, from (30) it follows at once that

$$B^2 = \frac{1}{1 - \frac{r^2}{3}(X_0 + 2X_1)} \tag{40}$$

or, by using (29)

$$B^2 = \frac{1}{1 + C_2 r^2 + 2r^2 \int \frac{X_1}{r} dr}\,; \tag{41}$$

where C_1, and C_2 are constants of integration.

As we can see from (39) and (41), this metric describes any static anisotropic sphere (see reference [12]) and can be expressed in terms of structure scalars X_1 and $Y_0 - Y_1$.

It is easy to check that the Schwarzschid interior solution corresponds to the case $X_1 = Y_1 = 0$. To illustrate the simplicity of the method, let us consider $X_0 = const$ and feeds this condition back into equation (38) to obtain

$$X_1 = \frac{C}{r^3} \tag{42}$$

IOP Conf. Series: Journal of Physics: Conf. Series **831** (2017) 012011 doi:10.1088/1742-6596/831/1/012011

Now, the regularity condition at $r=0$ via (40), implies $C=0$, then

$$-\mathcal{E}_1 - 4\pi\Pi_1 = 0 \Rightarrow \mathcal{E}_1 = \Pi_1 = 0\,. \tag{43}$$

Given that $\mathcal{E}_1(t,0)=0$, by the conditions of regularity at the origen, and $\mathcal{E}_1(t,R_\Sigma)=\frac{3M}{R_\Sigma^3}>0$ imposed by the boundary conditions, we demand that $\mathcal{E}_1(t,r)>0$. On the other hand Π_1 must be positive to guarantee the decreasing of radial pressure, via T.O.V. equation (See the reference [13])

Thus, we can see that necessarily the only static solution with homogeneous energy density, under the above considerations, is the Schwarzchild solution.

3.5.2. The non-static case

(i) **The Shear-free solutions** $\sigma_1=\sigma_2$
Under this condition, i.e., $\sigma=0$, the equation (17) turns out to be

$$\frac{\dot{B}}{B} = \frac{\dot{R}}{R} \tag{44}$$

Thus we have two options:
Option 1 If

$$R = bB \tag{45}$$

where b is an arbitrary function of r which can be taken as 1 without loss of generality, then we get

$$R = rB, \tag{46}$$

but it does not satisfy the regularity condition at $r=0$, nor the boundary condition.
Option 2

$$B = T\tilde{B} \quad and \quad R = T\tilde{R}, \tag{47}$$

where T is an arbitrary function of t and $\tilde{B}$, $\tilde{R}$ are arbitrary functions of r. Feeding back (47) into the equation (29), with $Z=0$, we get

$$\sigma = \sigma(t) \quad \Rightarrow \quad A = \frac{\dot{T}}{\sigma T}, \tag{48}$$

which implies that $a_1 = 0$. Therefore, we conclude that non-dissipative spherically symmetric shear-free fluids are necessarily geodesic.

(ii) **The Case** $\sigma_1 = 0$ **and** $Z=0$
If one assumes that $\sigma_1=0$ and that the fluid is non-dissipative then, the equation (29) can be integrated to obtain

$$\sigma_2 = \frac{\tilde{\sigma}(t)}{R}, \tag{49}$$

where $\tilde{\sigma}(t)$ is an arbitrary function of integration. Next, by demanding regularity of σ_2 at $r=0$, a fact obvious from (34), we have to put $\tilde{\sigma}=0$, which implies that $\sigma_2=0$, therefore we are lead to the static case.

(iii) **The Case** $\sigma_2=0$ **and** $Z=0$
If one assume that $\sigma_2=0$ and that the fluid is non-dissipative then, from (29) it follows at once that

$$J_1\sigma_1 = 0 \tag{50}$$

but, J_1 can not vanish, thus we get $\sigma_1=0$ i.e., static case.
From results obtained in ii) and iii) it follows that in the absence of dissipation ($Z=0$), using the regular center condition, if one of the functions B or R does not depend on t, the metric (6) is static.

(iv) **The Case $X_0 = X_0(t)$ and $Z = 0$** In this case, we obtain from the equation (38) that $X_1 = 0$ and the equations (35)-(37) become:

$$a_1(Y_1 - X_0 - Y_0) + 3J_1Y_1 = (Y_0 - Y_1)^{\dagger} \tag{51}$$

$$X_0^{\bullet} + (X_0 + Y_0 - Y_1)\sigma_1 + (X_0 + Y_0 + 2Y_1)\sigma_2 = 0 \tag{52}$$

$$X_0^{\bullet} + 2(X_0 + Y_0 - Y_1)\sigma_2 = 0 \tag{53}$$

Notice that if $X_0 = const.$, we get the static case, that we have analyzed before. Combining the equations (52) and (53) we find that

$$(Y_0 + X_0 - Y_1)(\sigma_1 - \sigma_2) + 3Y_1\sigma_2 = 0 \tag{54}$$

From the equation (54), it follows that

$$\sigma_1 - \sigma_2 = 0 \;\Leftrightarrow\; Y_1 = 0. \tag{55}$$

Again this leads us to the static case. In other words, shear-free and isotropic pressure conditions are equivalent, for non-dissipative fluids with homogeneous energy density.

4. Final Remark

We have shown, through very simple examples in the spherical case, how easily it is possible to obtain information from this Einstein-equivalent system and this encourage us to explore more general cases far from the spherically symmetric toy model we have developed in this work. The axisymmetric cases and also more general matter configurations are undergoing works that will be completed shortly.

Acknowledgments

J.O. and L.A.N. gratefully acknowledge the bilateral academic cooperation between Universidad de Salamanca and Universidad Industrial de Santander, which has promote this joint project. L.A.N thanks mobility funding from Vicerrectoría de Investigacin y Extensión, from Universidad Industrial de Santander

References

[1] Ellis G 1971 Relativistic cosmology in: Proceedings of the international school of physics "enrico fermi", course 47 *General Relativity and Cosmology* ed Sachs R (Academic Press, New York and London)
[2] Ellis G and Bruni M 1989 *Physical Review D* **40** 1804
[3] Ehlers J 1993 *General Relativity and Gravitation* **25** 1225
[4] Wainwright J and Ellis G F R 2005 *Dynamical systems in cosmology* (Cambridge University Press)
[5] Ellis G and Van Elst H 1999 Cosmological models *Theoretical and Observational Cosmology* (Springer) pp 1–116
[6] Tsagas C G, Challinor A and Maartens R 2008 *Physics Reports* **465** 61–147
[7] Ellis G, Maartens R and MacCallum M 2012 *Relativistic cosmology* (Cambridge University Press)
[8] Wald R M 2010 *General relativity* (University of Chicago press)
[9] García-Parrado A and Lobo G 2008 *EAS Publications Series* **30** 233–236
[10] Herrera L, Ospino J, Di Prisco A, Fuenmayor E and Troconis O 2009 *Physical Review D* **79** 064025
[11] Herrera L, Di Prisco A, Ibáñez J and Ospino J 2014 *Physical Review D* **89** 084034
[12] Herrera L, Ospino J and Di Prisco A 2008 *Physical Review D* **77** 027502
[13] Herrera L, Di Prisco A, Ospino J and Fuenmayor E 2001 *Journal of Mathematical Physics* **42** 2129–2143

IOP Conf. Series: Journal of Physics: Conf. Series **831** (2017) 012012 doi:10.1088/1742-6596/831/1/012012

La Hechicera Relativista

H. Rago

Escuela de Física, Universidad Industrial de Santander, Bucaramanga-Colombia and Departamento de Física, Universidad de Los Andes, Mérida Venezuela.

E-mail: rago@ula.ve

1. Un introduccin desde el exilio

Estas líneas deben oírse, están concebidas para el sonido, para ser dichas a veces como susurro, otras como grito. Por eso prefiero leerlas y equivocarme conscientemente y no en lugar de acertar improvisadamente. Son algunas divagaciones, disquisiciones, reflexiones sobre la física teórica en Mérida y Venezuela con una leves alusiones a los homenajeados en particular a Luis Herrera. Para contextualizar y como decía el famoso suplemento-comic El Fantasma, que algunos recordarán ¿'para los que llegaron tarde", haré un breve resumen de cómo nuestros homenajeados contribuyeron a cimentar la relatividad y la física teórica del otro lado de la vilipendiada frontera.

En una ocasión hace 10 años varios de los que hoy estamos presentes en esta sala nos reunimos en la Isla de Coche, al oriente del país que se llamó Venezuela, para un homenaje a Luis Gaucho Herrera Cometta, en ocasión de sus jóvenes sesenta años.

Hoy, nos reúne lo que podríamos llamar la versión 2.0 de aquel III Taller de relatividad y astrofísica relativista, esta vez en homenaje a los 140 años equitativamente repartidos, de Luis Herrera y Rodolfo Gambini. Segunda edición ampliada, corregida y aumentada y donde constataremos que si veinte años no es nada, en su justa mitad es bastante y muchas cosas pueden contarse. De modo que lo que sigue es un remake que no excluye el autoplagio y con una epílogo que actualiza lo ocurrido en el interín.

El comité organizador en aquel entonces, al que pertenecí junto con Luis Núñez, Willians Barreto, Alicia Diprisco, Mario Cosenza, Pío Arias, María Esculpi y Ernesto Fuenmayor, supo brindar una calurosa –en las distintas acepciones y temperaturas de la palabra– bienvenida al patio y sobre todo a los invitados extranjeros.

Hoy toca agradecer al comité organizador de este encuentro, Antonio C. Gutierréz-Piñeres (Universidad Tecnológica de Bolívar), Edison Montoya (Universidad Industrial de Santander), Jorge Muñiz (Universidad Tecnológica de Bolívar) y Luis A. Núñez (Universidad Industrial de Santander) tener el privilegio y la posibilidad esta vez como extranjero en esta tierra, de conversar, compartir y seguir aprendiendo hoy y aquí, de nuestros homenajeados.

En aquella ocasión quiso el comité organizador que yo hablara a cuenta de pionero o más bien, de viejo, que viene siendo lo mismo, del desarrollo de la relatividad, en la ciudad de Mérida, es decir, en la Universidad de Los Andes, y por extensión, de la física relativista en Venezuela.

La Hechicera relativista es una acertada metáfora que debo a Willians Barreto. La Hechicera es el hermoso lugar donde funciona la facultad de ciencias de la ULA?

Se trata de hilar en forma de recuerdos deshilvanados, algunas nostalgias, algunas fechas, algunos esfuerzos y algunos números en una breve historia, con protagonistas, co-protagonistas, contrafiguras y naturalmente, donde no puede faltar el muchacho de la película (ya no tan muchacho, como todos sabemos), y esta vez diez años más joven.

La moraleja de este corto resumen será paisaje usual para muchos de los presentes. Al fin y al cabo, todo el que haya sido pionero en tierras latinoamericanas de alguna aventura científica, sabe o por lo menos intuye, que el esfuerzo tiene que ver con cómo conciliar tanto páramo con el tensor de Riemann-Christoffel; tanto mastranto y paisaje rupestre, con la formulación de loops de la teoría cuántica de la gravitación y tanta incomprensión oficial con la parte magnética del tensor de Weyl.

2. Tres Fases y un destino

Pido de antemano disculpas por comenzar autobiografiando y por repetir autoreferencias, pero, no queda más remedio. Una genética manía didáctica me lleva a considerar tres fases:

- FASE 1. En 1971, cuando Cygnus X1 fue identificado como un candidato fuerte a agujero negro y finalizando mi carrera en la UCV de mis tormentos, un allanamiento militar y clausura hasta nuevo aviso, de la Universidad, nos empujó atropelladamente y afortunadamente (nunca he agradecido tanto un allanamiento) a tierras merideñas, donde a la sazón otros soñadores habían recién fundado un departamento de física en una incipiente Facultad de Ciencias en la Universidad de Los Andes.
  No creo exagerar si dijera que el contexto académico era desolador. Para contextualizar e imaginarnos el desierto, reseñemos que en el Consejo universitario de la época se debatía y criticaba a la Facultad de Ciencias por sus pretensiones de construir una biblioteca que incluía libros y revistas, claro, en inglés. óigase bien: en inglés... ¿Cuál es el idioma de este país? preguntaban nuestros nacionalistas consejeros hace cuarenta y tantos años. La relatividad de Galileo podía ser un desatino o en todo caso una oscuridad distante. No te cuento la de Einstein. Hawking estaba por publicar su contribución acerca de la radiación térmica en el horizonte de agujeros negros, mientras un servidor dedicaba amplios y vanos esfuerzos para entender los misterios de un críptico tomo II, Teoría Clásica de Campos y Relatividad, de Landau, sin nadie a quien comentarle las veleidades de la métrica espaciotemporal, tan esquiva y tan riemanniana ella, que no me concedía ni uno sóla de sus curvaturas.
  Una maestría en el IVIC, en Caracas, me acercó a uno de los jurados de mi tesis de grado. Era Luis Herrera Cometta en lo sucesivo LHC, siglas que naturalmente deben confundirse con las del Large Hadron Collider. Fue el inicio de una colaboración persistente, que me permitía al menos disponer de un interlocutor a 700 km de distancia, pero capaz de aclarar dudas elementales y profundizar las fundamentales, como debe ser. Este inicio produjo una transición de fase que nos condujo a la 2 Fase.
- FASE 2 (*circa 1979*) Coincidiendo con la aparición del modelo inflacionario en cosmología y con el descubrimiento del primer quasar doble 0957+561, y producto de la colaboración con LHC, se abrió la compuerta a una serie de trabajos coautoreados entre Mérida y Caracas, entre la ULA y la UCV. Ello significó tesis, nuevos licenciados, hoy muchos de ellos solventes investigadores independientes, incluso algunos ya pensionados.
  El momento coincide con la incorporación al departamento del joven Luis Núñez (todos entonces lo éramos), de Alberto Patiño, de Nelson Pantoja, Adel Khoudeir, más tarde Alejandra Melfo, Héctor Hernández, José Humberto Ocariz, Umberto Percoco, algo después, William Barreto y otros. Más tesis, doctorados, muchos bajo la lupa admonitoria de LHC. Se respiraba efervescencia, algo bueno estaba pasando. De los 10 doctores (para la época) tutoreados por nuestro homenajeado, cuatro estábamos en la ULA.

Enseñando nuestras dudas las disolvíamos. Surgieron seminarios periódicos, entrecruzamiento de líneas de investigación, materias electivas en el área, aprendimos a contactar a colegas de sitios distantes. Jaume Carot puede atestiguarlo, Sergio Hojmann y Diego Pavón también, José María Ibáñez, la conexión gallega en pleno. Hubo tradición, hubo aprendizaje con algo de artesanal, en encuentros cercanos de todo tipo entre profesores y alumnos. Aprendimos a dialogar con quienes en otras universidades del país invertían idénticos esfuerzos con idénticos propósitos y tal vez con idénticos resultados. El proceso de replicación de relativistas, estaba en marcha. La masa crítica se había obtenido. La fase solitaria de onanismo extremo cedía lugar a la más atractiva lujuria de la práctica orgiástica y el disfrute colectivo. Nueva transición de fase y llegamos así a la tercera.

- Fase 3.- (*circa 1992*) EL AUGE. Tres eventos importantes merecen ser reseñados durante esta tercera fase. La línea de Relatividad numérica, la Escuela ?La Hechicera? en Relatividad, Campos y Astrofísica, y la fundación del Centro de Física Fundamental y el postgrado en Física Fundamental, cariñosamente nombrado como Pofifu.
  La maduración de una línea sesgada hacia la relatividad numérica fue la consecuencia de la colaboración entre LHC y Luis Núñez con los trabajos de su tesis doctoral y sus derivaciones. Posteriormente esta línea se nutriría eficazmente por quien al otro lado del país hacía formidables esfuerzos por no arar en el mar caribe cumanés, Willians Barreto. Esto significó intercambio y colaboraciones con gente como Joan Masso, Ed Seidel, Jefrey Winicour y otros. En una apuesta por el atrevimiento, diría que de manera indirecta convirtió a la ULA en pionera en asuntos de informática, correo electrónico, conexión a Internet; y más indirectamente aún, convirtió a Mérida en foco de irradiación informática al resto del país. Mérida era en esos días la ciudad con más cyber-cafés por habitante en el país y en América Latina. A un informe de la UNESCO como prueba me remito. No es injusto verlo como un subproducto del desarrollo de la relatividad en Mérida.
  Lejanos lucen ya los tiempos en los que debíamos ir a la biblioteca a consultar un semanario llamado Current Content, impreso en papel como de periódico y cuya primera mitad eran fotocopias de los índices de las principales revistas, y la segunda, las direcciones de los autores que aparecían en el índice. Te llamaba la atención el título de un trabajo, buscabas la dirección del hindú autor del trabajo. Siempre era un hindú. O un pakistaní. ¿trabajábamos un área tercermundista de la física? Le escribías en una postal preimpresa solicitándole una copia de su trabajo tal y tal, publicado en la revista tal y tal. Cuatro meses después, con buena suerte, te llegaba el trabajo para constatar, con mala suerte que realmente no tenía nada que ver con el tuyo.
  Los tiempos cambiaron, definitivamente y los jóvenes de hoy (no me refiero a los homenajeados) jamás entenderán que hubo una época en la que no bastaba hacer click para bajar un artículo, y debías esperar meses para tenerlo.
  En 1992, en febrero y en noviembre recibimos dos pésimas noticias disfrazadas de golpes de estado, afortunadamente fracasados. No pasaría demasiado tiempo antes de que vislumbráramos las nefastas consecuencias que habrían de tener en la vida del país y en la vida académica.

3. La Escuela La Hechicera

Pecaré de inmodesto al afirmar que la estabilización a partir del 1995, de la escuela La Hechicera en relatividad, campos y astrofísica, fue importante en el microcosmos vinculado a la relatividad y a la física teórica en el país y tal vez algo más allá. Reunirnos ritualmente año tras año, noviembre tras noviembre, en nuestra propia versión del mito del eterno retorno, jóvenes investigadores, veteranos profesores, ávidos estudiantes, a caernos a pasiones mutuas, a enseñar y a aprender, y a aprender a enseñar y enseñar a aprender, fue una formidable experiencia. Ambos muchachos de la película de hoy fueron profesores de la escuela y en muchas circunstancias

IOP Conf. Series: Journal of Physics: Conf. Series **831** (2017) 012012 doi:10.1088/1742-6596/831/1/012012

sirvieron de guía, sugiriendo profesores, o contactando colegas. Permítanme mencionar a algunos de los profesores que tuvimos en las escuelas:

(i) Escuela: Noviembre1995, dedicada a Carlos Aragone, Ponentes: Rodolfo Gambini, Joan Masso y Héctor Rago.

(ii) Escuela: Noviembre 1996, Profesores Goran Senjanovic, Juan Jiménez y Luis Herrera Cometta.

(iii) Escuela Noviembre 1997. Profesores Nilton Santos, Fernando Quevedo y José Franco

(iv) Escuela, Noviembre 1998 , José Luis Ballester. Enric Verdaguer y César Gómez.

(v) Escuela, Noviembre 1999, Profesores Ana María Font, Jesús Gallego y Jorge Zanelli.

(vi) Escuela, noviembre 2000, Gustavo Bruzual, Esteban Roulet, Juan García-Bellido.

(vii) Escuela, noviembre, 2001 Nuria Calvet, Miguel Alcubierre, Tomás Ortín.

(viii) Escuela, noviembre del 2002, Jaume Garriga, Alex Pomarol y Matías Zaldarriaga.

La escuela La Hechicera representó una treintena de cursos de alto nivel académico, trescientas horas de clases impartidas a seiscientos alumnos de estos paisajes latinoamericanos e hispanos, materializados en alrededor de 2000 páginas publicadas en 8 libros que en las bibliotecas y en la web, donde están colocados, multiplican su efecto benéfico. Fue grato recientemente oír de profesores de la UniAndes de Bogotá, hoy reputados investigadores, lo importante que fue en la formación de la astrofísica colombiana, la escuela La Hechicera en relatividad, campos y astrofísica.

Hoy el destino de la Escuela es una punzada en la esquina más académica de nuestro hígado de relativista. Desencuentros con la cada vez más politizada e intolerante burocracia oficial, nunca logramos financiamiento, y tras ocho ediciones consecutivas, la escuela La Hechicera en Relatividad, Campos y Astrofísica se interrupió definitivamente en el año 2002.

Digno de mención son los seminarios de Relatividad de Caracas (IVIC, UCV, USB) con participación relevante de LHC y de Rodolfo, junto con su álter ego en asuntos científicos, Antonio Trías, un dúo tan exitoso en física como el Lennon- Mccartney en música, o Tom Jobim y Vinicius da Moraes, por mantener el tono tropical. Estos seminarios marcaron la relatividad y la física teórica en el centro del país, así como la Escuela la Hechicera lo hizo en el occidente. Los talleres como este tercer taller de relatividad y astrofísica relativista son también un ejemplo de actividad.

Pero más allá de los números y sus enumeraciones, quisiera volver a la nostalgia. Es obvio que la situación ha cambiado en estos treinta años desde cuando comenzamos en estos ajetreos. Los alumnos de pre y postgrado disponen de profesores que los guían por los vericuetos del espaciotiempo, a veces curvo y otras no. Hay interlocutores, se ha rutinizado en el buen sentido de la palabra la enseñanza de la relatividad y sus primas cercanas. A golpes a veces bajos, y otras veces menos traumáticos, aprendimos el oficio.

Entre balbuceos y moretones fuimos descubriendo por ensayo y terror, los mecanismos ocultos de la profesión. Descubrimos aterrizando en la realidad que lo que cuando estudiantes presentíamos como una romántica utopía de arrancarle a dentelladas los secretos a la naturaleza, podía parecerse más a esta maldita integral que no sale, el programa que no corre, el árbitro necio que no entiende; que además tengo que llenar tres informes finales, y cuatro solicitudes de financiamiento con cronograma, metodología, hipótesis de trabajo y otros extravíos. Pero aprendimos a hacerlo, y a veces hasta logrando transmitirle a los alumnos, la pasión por lo que hacemos. Como el atribulado personaje de"El día que me quieras" que quería ver la historia y terminaba siempre por oírla, en la luminosa descripción de José Ignacio Cabrujas, nuestra gran unanimidad en asuntos de teatro, (y me perdonan de nuevo el localismo), nosotros también quisimos ver la historia y no sólo oírla. Fue la época del auge. Como la canción de Sinatra¿'it was a good year?"

IOP Conf. Series: Journal of Physics: Conf. Series **83**1 (2017) 012012 doi:10.1088/1742-6596/831/1/012012

En un ejemplo de acción a distancia de efectos visibles y contundentes, quisiéramos reconocerle a LHC haber estado allí, como los buenos árbitros de futbol, sin sentirse demasiado, pero llevando el partido con mano experimentada. No hay que menospreciar el efecto del ritual del ponche en el Hotel Valle Grande de Mérida, ni las comilonas con película incluida, que remataban cada visita de LHC a nuestra universidad, en trance de ir adquiriendo una cultura de cómo funcionan los engranajes de la ciencia, las relaciones entre pares, el ardid de la publicación en revistas, el montaje de reuniones de trabajo, seminarios y demás hábitos de la rutina científica. Enseñándonos a no cambiar camino real por vereda, LHC *dixit*. Siempre queriendo ir a Mérida pero siempre queriendo irse lo más rápido posible luego de cumplido los rituales de trabajo que incluían ponerse al día tras la práctica de una correcta metodología chismográfica, de los dimes y diretes, comentarios maledicentes y chismes varios de la comunidad de relativistas.

4. A manera de epílogo

Entre un homenaje y otro median diez años cronológicos. Pero sabemos que el tiempo es relativo. O es una ilusión, aunque persistente. Sabemos también que ambas frases las pronunció la misma persona. Durante este interín hubo algunas noticias en la física.

(i) En 2009 se lanzó el Planck para escrutar las fluctuaciones de la radiación cósmica con una resolución sin precedentes. Fue apagado hace tres años y sus datos nos seguirán informando cada vez mejor del universo temprano.

(ii) En el 2011 a los neutrinos del experimento OPERA en Italia le dio por ser transgresores y empezaron tener ínfulas de superlumínicos, y amenazaban con socavar nuestra fe y hacernos dudar del dogma.

(iii) 8 de junio de 2012, Opera anuncia que lograron reducir a los neutrinos transgresores y los obligaron a viajar a velocidades políticamente y físicamente correctas. Y los experimentadores aprendieron a enchufar los cables correctamente. Y todo volvió a la normalidad

(iv) El 4 de julio de 2012 el CERN anuncia que LHC detecta una nueva partícula que luce como el bosón de Higgs, huele como el bosón de Higgs y sabe al bosón de Higgs. Lo llamaron el bosón de Higgs porque el nombre de ?la partícula de dios? fue declarado persona non grata por los físicos.

(v) En el 2014 Bicep logra detectar ondas gravitacionales primordiales donde sólo había polvo.

(vi) En el 2015 Bicep-Plank no logran detectar nada. Todo vuelve a la normalidad.

(vii) 15 de septiembre de hace un año, los dos detectores de LIGO se estremecen un milésima parte del diámetro de un protón. El estremecimiento en las redes sociales es mucho mayor: el chisme circula por los stalkers de tuiter: ¿se detectaron ondas gravitacionales?

(viii) 25 de Noviembre se celebra el centenario de las ecuaciones que Einstein diseñó para entender más y mejor a la gravedad. Y todo vuelve a la normalidad.

(ix) 12 de febrero de 2016 el director ejecutivo del LIGO David Reitze economizando palabras dijo lacónicamente: "Ladies and gentlemen, we have detected gravitational waves. We did it!". Agujeros negros en coalescencia emitieron hace1,3 millardos de años, en fracciones de segundo mucho más energía en ondas gravitacionales que todas las estrellas del universo observable. Los stalkers estaban en lo cierto y las redes sociales volvieron a estallar. Einstein sonrió y dijo más lacónicamente "I did it"

5. La Caída

Mientras tanto del otro lado de la frontera nada vuelve a la normalidad. Las noticias del terreno no son nada halagadoras.

IOP Conf. Series: Journal of Physics: Conf. Series **831** (2017) 012012 doi:10.1088/1742-6596/831/1/012012

En apenas diez años duele constatar el profundo y lascerante retroceso. Las asonadas militares tuvieron consecuencias. Nefastas, sí. Previsibles, sí. El esfuerzo de treinta y tantos años puede disolverse en nada. El tiempo es una ilusión, ya lo dijimos. El país también puede ser un extravío. El homenajeado de hace diez años está en Salamanca tan activo como siempre. Luis Núñez incansable como nunca, en Bucaramanga, Alejandra Melfo entre Holanda e Italia, William Barreto en Brasil primero y en algún otro lugar después. Umberto Percoco prepara pizzas, el único de los doctores jóvenes de reciente ingreso, en tres meses vivirá en Uruguay, un servidor intenta agradecido en esta tierra acogedora, un esfuerzo en divulgación sin morir en el invento. Lo demás languidece, lo demás es una anomia. La universidad también puede ser una asfixia.

Se nos dirá que la ciencia es universal y que no importa dónde se trabaje, es cierto, pero una generación de nuestros estudiantes, alrededor de diez, casi rozan el bosón de Higgs en el CERN, y fueron privilegiados. No oyeron la historia, la hicieron. Y es un orgullo indeleble para ellos y para los que los formamos. Eso no volverá a ocurrir porque el 30% de la academia universitaria está afuera del país. Los fascismos no se las llevan bien con las universidades.

Si alguna vez el flujo de estudiantes era de estas tierras colombianas hacia el país que intentábamos construir, en estos momentos de reflujo hay que agradecer tantos brazos abiertos en este país que busca desesperadamente la paz.

Los venezolanos del lado de acá y del de allá de la frontera hubiésemos querido que un homenaje a Rodolfo Gambini y a Luis Herrera se hiciera con todos ustedes, pero en Venezuela. Ellos lo merecen, nosotros también. Pero los venezolanos a ambos lados de la frontera sabemos, nos consta, juramos?que eso es imposible. De ese tamaño es la desolación y de ese tamaño es el agradecimiento por poder realizar el homenaje hoy y aquí.

Pido de nuevo disculpas por el quejumbroso tono de lamento. Y volvamos al tema que nos ocupa: Nos consta la urticaria ontológica que siente LHC en las cercanías de homenajes formales con olor a naftalina. Nos consta que su herejía y su irreverencia congénita se irritan ante los discursos laudatorios. Quisiéramos evitarle la urticaria, las irritaciones y tanta alergia. Evitaremos la pompa y la naftalina. No evitaremos en cambio, el afecto en trance de agradecerle los favores recibidos.

Luis y Rodolfo, lo baila'o no nos lo quita nadie y sin contar lo que nos falta por bailar. Además dentro de diez años en el próximo homenaje habrá mejores noticias de la comarca criolla. Simplemente Sí es algo personal: la física teórica en Mérida y en Venezuela y mucho más allá, les deben una. Yo modestamente cumplo con pagar mi cuota y lo hago en forma de un caluroso fuerte y persistente abrazo.

Muchas gracias.